AF327480

Quantum Chemical Aspects of Cationic Polymerization of Olefins

Quantum Chemical Aspects of Cationic Polymerization of Olefins

V.A. Babkin

G.E. Zaikov

K.S. Minsker

NOVA SCIENCE PUBLISHERS, INC.
Commack, NY

Editorial Production: Susan Boriotti
Assistant Vice President/Art Director: Maria Ester Hawrys
Office Manager: Annette Hellinger
Graphics: Frank Grucci
Acquisitions Editor: Tatiana Shohov
Book Production: Ludmila Kwartiroff, Christine Mathosian,
                Joanne Metal and Tammy Sauter
Circulation: Iyatunde Abdullah, Sharon Britton, and Cathy DeGregory

Library of Congress Cataloging-in-Publication Data
available upon request

ISBN 1-56072-501-X

Copyright © 1997 by Nova Science Publishers, Inc.
                6080 Jericho Turnpike, Suite 207
                Commack, New York 11725
                Tele. 516-499-3103   Fax 516-499-3146
                E-Mail: Novascience@earthlink.net

If you want to outlast yourself and be in respect with your posterity, leave after virtuous family and a good book.

Pythagor,
Ancient Greek
mathematician and philosopher.

CONTENTS

Preface   ix

Introduction   xi

Chapter 1. The bases of cationic polymerization.   1
    1.1   Common Information   1

Chapter 2. Catalytic activity of Lewis and Brenstedt complex acids as active centers of cationic polymerization of olefins.   9
    2.1.   Initiation of cationic polymerization of olefins by complexes of Lewis and Brenstedt acids.   9
    2.2.   Characterization of active centers of complex acids of Lewis and Brenstedt in the processes of cationic polymerization of olefins.   17

Chapter 3. Quantum chemical methods of calculation of the complexes as initiators of cationic polymerization of olefins.   21
    3.1.   Characterization of semiempyric quantum chemical methods of calculation.   21
    3.1.1. CNDO/2 method.   23
    3.1.2. MiNDO/3 method.   39

Chapter 4. Quantum chemical calculation of complex Lewis and Brenstedt acids as the active centers of cationic polymerization of olefins.   43
    4.1.   Complexes of hydrogen chloride with aluminum chlorides.   48
    4.2.   Complexes of hydrogen fluoride with boron fluorides.   53
    4.3.   Complexes of alkaline metals.   54
    4.4.   Aquacomplexes of aluminum chlorides.   60
    4.5.   Aquacomplexes of magnesium chlorides.   83
    4.6.   Aquacomplexes of boron fluorides.   91
    4.7.   Complexes of alcohol with aluminum chlorides.   93
    4.8.   Complexes of alcohol with magnesium chlorides.   109
    4.9.   Complexes of alcohol with boron fluorides.   112
    4.10. Correlation of proton donor and carbcationic activity of alcohol of the complexes of Lewis and Brenstedt acids.   115

Chapter 5. Applied quantum chemistry of cationic polymerization of olefins.   125
    5.1.   Indexes of reactionary ability of cationic polymerization of olefins.   125
    5.2.   Cationic polymerization of olefins as the coordinated process.   126
    5.3.   Interconnection of acidic catalytic properties of complex acids of Lewis and Brenstedt.   129

References.   133

Index.   137

PREFACE

The monograph represents an attempt to generalize the material published on electron structure, geometry and behavior of complex Lewis and Brenstedt acids during initiation of cationic polymerization of olefins.

The results of calculations of electron-energetic characteristics, performed according to quantum chemistry methods, are the basis for the explanation of acidic catalytic properties of complex electrophilic catalysts and clarification of the mechanism of elementary acts of the polymerization processes. This may promote the research of new more effective and selective catalysts. Questions are broached about of the nature of active centres (AC) and the role of counterion during cationic polymerization of olefins.

The authors would like to thank: Corresponding Member of A.S. Belorus Doctor of Chemistry Ju.A. Sangalov, Doctor of Physics and Mathematics O.A. Ponomarev, Candidates of Chemistry V.M. Junborisov and M.M. Karpasas for fruitful discussions and help in the exploration of the methods of quantum chemistry and their application to cationic polymerization of olefins. We also want to thank Doctor of Chemistry Ju.A. Prochukhan and Candidate of Chemistry Ju.Ja. Nelkenbaum for the performance of experimental part of the investigation, and also Candidate of Chemistry M.I. Arcys for the help in manuscript preparation.

Doctor V.A. Babkin, Professor K.S. Minsker
Bashkirsky State University, Ufa, Bashkortostan, Russia
Professor G.E. Zaikov
Institute of Chemical Physics RAS, Moscow, Russia

INTRODUCTION

Cationic polymerization, proceeding in the presence of electrophilic catalysts, is one of the main methods of obtaining polymeric products. Cationic polymerization was discovered by Deville in 1939. Apparently, systematized investigations of this process began in 20s of XX century from the works of Shtaudinger [1].

Cationic polymerization is the chain process, proceeding in three stages: 1) initiation – the formation of active centres, carrying positive charge; 2) chain growth – attachment of a monomer to the active center; 3) restriction of the chain growth – the break of the material and (or) kinetic chain. Meanwhile, each of these stages may include often more than one elementary act. This circumstance as well as high sensitivity of the process to the traces of foreign matter, very high rates of the growth reaction, in many cases flare front of the reaction proceeding and nonisothermal character of the process, makes difficult the obtaining of generalized information about cationic polymerization, and as a consequence the creation of unique, including quantitative level, general strict theory.

The complexity of the situation does not exclude, but supposes and stimulates fundamental investigations in the branch of thin mechanism of cationic polymerization. In many cases sufficiently new information may be obtained only applying quantum-chemical calculations.

The monograph considers the achievements in the branch of the study of the process mechanism (initiation and chain growth, principally) during cationic polymerization of olefins in particular, reaction trajectory, the role of counterion, etc. on the example of the application of metals halogenides with proton donors as catalysts of the complexes. These binary compounds are traditional acidic initiators of the reactions of homo- and copolymerization of olefins, alkylation, etc. [2, 3]. The change of the nature of a proton donor (H_2O, HCl, ROH, etc.) ensures formation of complexes, possessing a wide range of activity and action selectivity, even at the application of one and the same metal halogenide. However labile character of complexes leads in general case to the existence of a selection of compounds in the system, including active centers differing by the composition as well as by the catalytic properties, in dependence on the method of storage and production of the catalyst, performance of catalytic polymerizing process, the nature of the substrate and other factors. Naturally this makes the analysis of the acidic and catalytic properties difficult and as a consequence the validity of active application of one or another system of the catalysts.

One of the basic characteristics which defines the catalytic properties of complex compounds is the acid strength. As there are no reliable experimental methods available for determination of complex catalysts acidity, H-complexes in particular [3], there appears the important meaning of theoretical calculation methods for estimation of the parameters of acid strength, correlating with the values of universal index of acidity pKa [4]. The development of chemistry and physical chemistry of complex compounds of halogenides (alkyl halogenides) of aluminum and other elements with the proton donors, creation of the models, similar to various cases of the complex formation, programming support of quantum-chemical calculations, etc. allow to estimate the reliability of alternative schemes of initiation, inaccessible for direct experimental confirmation, the role of counterion which is still "the thing inside itself" according to the theoretical point, to obtain the information about the dynamics and the mechanism of the catalytic process. The correlations set of calculated and experimental data obtained for a single type of the complex compounds make the prognosis of new complexes as potential catalysts possible. Consequently, beside the development of the concept about the role of the acid strength in behavior of the catalysts, selection of initiators for particular processes and directly the methodology of the performance with the catalysts becomes more confirmed. The book shows the data on quantum-chemical calculation of the complexes of aluminum (magnesium) chlorides (alkyl chlorides) and boron fluorides (alkyl fluorides) with various proton donors as identification of the active centers of the systems and clarification of the role of counterion and monomer on the stages of initiation and chain growth during cationic polymerization of olefins.

CHAPTER 1

THE BASES OF CATIONIC POLYMERIZATION

1.1. Common information

Cationic polymerization of olefins is the process in which active growing particles carries the positive charge. Usually all cases of the formation of a polymer under the influence of cationic catalysts, possessing electrophilic (acidic) character, are corresponded to this class of reactions.

Systematized investigations of cationic polymerization, began even in the 20s of the present century, in further time were deeply developed in the works by S.S. Medvedev, N.S. Enikolopov, V.A. Kabanov, B.L. Ierusalimsky, V.P. Zubov, S.S. Skorokhodov (USSR), J. Kennedy (USA), P. Plesh, S. Pasinkevitch (Poland), J. Furukava (Japan) and others.

Cationic polymerization is initiated by a wide range of substances which are electron acceptors – Lewis acids, in particular:

a) protonic acids – H_2SO_4, H_3PO_4, $HClO_4$, HIO_4, $HClSO_3$, HCl, HF, $HFSO_3$, CF_3COOH, $CClCOOH_3$, etc.;

b) aprotic acids (general form is MeX_n, where Me is a metal, X is a halogen); the most widely spread catalysts from this group are BF_3, BCl_3, $AlCl_3$, $AlBr_3$, $SnCl_4$, $SnBr$, $TiCl_4$, $SbCl_3$, $SbCl_5$, $ZnCl_2$, $HgCl_2$, $FeCl_3$, and other mixed halogenides of metals also, for example $AlBr_3 \cdot MeX_n$, $AlI_3 \cdot MeX_n$, etc.;

c) halogens – I_2, ICl, IBr;

d) Carbonium salts – for example Ph_3, C^+, $C_7H^+_7$, A^-, where A is Cl, $SbCl_6$, PF_6 etc.;

e) oxonium salts – for example R_3O^+, A^-, where A is BF_4, $SbCl_6$, and R is CH_3, C_2H_5, etc. Cationic polymerization is also initiated by high energy radiations, natural and synthetic alumosilicates, ceolytes, etc.

Cationic polymerization is characterized by the leading of the compounds providing the highest rates to the formation of polymers with the highest molecular mass. In this case quantitative comparison of the catalysts activity of various groups is difficult, because the mechanism of their action in various systems may be different. Activity of protonic acids depends on their acidic strength. Weak acids do not initiate the process.

Chlorides of metals of III and IV group of Mendeleev's periodic system are sufficiently active in the group of aprotic acids usually. Acidity of these compounds is the higher, the lower electrically negative the halogen atom is. Acidity degree of aprotic acids during cationic polymerization performance depends on the nature of the monomer and may change according to the change of the conditions of the reaction (temperature, pressure, medium polarity). For typical monomer – isobutylene – activity of the catalysts of this (at -78°C) decreases in the list:

$$BF_3 > AlBr_3 > TiCl_4 > TiBr_4 > SnCl_4 > BCl_3 > BBr_3$$

Initiation of cationic polymerization by aprotic acids requires the presence of a t least sufficiently small (up to 1.1 mmole/mole) amounts of another ionogenic substance – cocatalyst (promoter) beside the main catalyst (Lewis acid, for example). These cocatalysts were discovered by Ipatiev and Gross in 1936. The role of these substances can be played by water, hydrohalogenic acids of HX type, alkylene oxides, α-, β-halogenesters, lactanes, etc. It is admitted, that the presence of the cocatalyst is always required for cationic polymerization of hydrocarbonic monomers, initiated by aprotic acids.

A wide range of organic substances is polymerized under the influence of the catalysts of cationic type: a) nonsaturated compounds, forming polymers by means of scission of double and triple bonds; b) cyclic compounds, forming linear polymers by means of the cycle break. The following compounds relate principally to the first class: α-olefins, some polyenes, oliphatic aldehydes and ketones, thio-aldehydes and thuo-ketones, vinylic esters, ketenes, isocyates, cyanamydes, etc. The following compounds should be mentioned among the substances of the second class: cyclopropen, simple cyclic ethers, thio-ethers, cyclic formalies and acetalies, lactones, lactomes, pyridin, quitones, etc.

Activity of monomers is the kinetic characteristic defining the rate of attachment of the monomer to the active center and depending on the affinity of disclosing double or triple bond of the monomer to growing cation. The affinity of aliquot bond to cation (or its nucleophility) grows with the increase of the electron density in it. In its turn this depends on the nature and disposition of the substitutors. Electrodonating substitutors, for example, alkylic and alkoxile, increase electrophility of aliquot bond, and electroacceptor ones – decrease.

Electron density on aliquot bonds at cationic polymerization influences the activity of the monomer sufficiently than spatial factors.

In cationic polymerization 1,1-disubstituted olefins with electrodonating substitutors are the most active among olefins. Activity of the monomers, polymerizing with disclosure of the cycle increases with the increase of their strain.

Substitution in the cycle induces the decrease of activity. Activity of simple cyclic ethers decreases with the growth of the cycle size. Nonsaturated cyclic esters, for example, dihydrofuranes, are polymerized by double bond.

Great role at cationic polymerization is played by the medium. Two effects are characteristic: a) stabilization of forming charged particles; b) the change of the active centers reactivity.

Stabilization by means of the medium is required at the occurrence of active centers of the process, because in this case energy losses for the hydrolysis of chemical bonds a t the formation of initiating ions are compensated. The change of reactivity of AC in various media proceeds by means of:

a) the influence of the polar medium;
b) the influence of its cocatalytic action;
c) the influence of specific solvatation;
d) the formation of complexes with the components of the system.

It is accepted in cationic polymerization that among these factors (excluding the influence of the medium as catalyst) the predominant one is the medium polarity. In this case it is traced the following tendency: at the increase of the medium polarity the rate of the process and molecular mass of forming polymer increases.

The character of the dependence of the rates of different elementary stages proceeding at cationic polymerization is defined by polarity of the medium and the nature of interacting particles (ions, polar and nonpolar molecules). The increase of polarity of the medium usually increases the rate of initiation and decreases the rate of the chain break. It is difficult to say something concrete about the stage of the chain growth, because the increase of the mean distance between the components of ionic pair must lead to the increase of the rate constant of the growth. However, the degree of the reactivity change of ionic pairs in dependence on the distance between cation and ion is not known yet.

Sufficient influence on the kinetics of cationic polymerization is performed by solvating ability of the solvent. For example molecules of the solvent, able to form complexes with the molecules of the catalyst or monomer, may change sufficiently its activity, and in some cases can suppress it completely. Another example of the influence of solvation by the solvent is presented by polymerization in media of aromatic hydrocarbons, which are the donors of electrons and in this connection are able to solvate the molecules of the acceptors of electrons – the catalysts of cationic polymerization. It is characteristic that paraffinic hydrocarbons and benzene possess similar dielectric penetrability. However, cationic polymerization in benzene proceeds as a rule with higher rate, than in paraffinic hydrocarbons because of better solvation of the growing particles.

It should be mentioned particularly that cationic polymerization is sufficiently sensitive to the small amounts of admixtures, that stipulates the difficulty of obtaining reproductive experimental results. It can be often observed that at low concentrations of the admixtures the increase of the general rate of the process is obtained, which is connected usually with its cocatalytic performance. The increase of admixtures content in the reaction zone leads to the upside down effect, stipulated by the reactions of the restriction of the chain growth [5].

In general case cationic polymerization is the chain process, representing often sufficiently complicated reaction. These circumstances as well as high sensitivity of the process to the traces of the foreign matter makes the obtaining of the information on the mechanism of cationic polymerization difficult. In this connection it may be pointed out the following sequence of the reactions for the realization of cationic polymerization:

$$\text{I. Initiation.}$$
$$H^+, A^- + CH_2=CHR \rightarrow CH_3-C^+HR, A^-$$
$$R^+, A^- + CH_2=CHR \rightarrow R-CH_2-C^+HR, A^-$$
$$K^+ + A^- + CH_2=CHR \rightarrow K-CH_2-C^+HR + A^-$$

$$\text{II. Chain growth.}$$
$$\sim CH_2-C^+HR, A^- + CH_2=CHR \rightarrow \sim CH_2-CHR-CH_2-C^+HR, A^-$$

$$\text{III. Limitation of the chain growth.}$$
$$\text{a) chain break}$$
$$\sim M^+, A^- \rightarrow \sim MA$$
$$\sim M^+, BA^- \rightarrow \sim MB + A$$
$$\sim M^+, A^- + X^- \rightarrow \sim MX + A^-$$

$$\sim CH_2 - C^+, A^- \rightarrow \begin{array}{l} CH_3 \\ \\ CH_3 \end{array} \quad \begin{array}{l} \sim CH_2 - C = CH_2 + HA \\ \quad\quad CH_3 \\ CH = C(CH_3)_2 + HA \end{array}$$

b) chain transfer

$$\sim CH_2\text{-}C^+HR,\ A^- + CH_2 = CHR \rightarrow\!\!\!/ \sim CH = CHR + CH_3\text{-}C^+HR,\ A^-$$
$$\sim CH_2\text{-}C^+HR,\ A^- + RX \rightarrow\!\!\!/ \sim CH_2\text{-}CHRX + R^+.$$

Cationic polymerization of olefins is exothermal reaction, because it is connected with the transformation of π-bonds into G-bonds [6]. The heat of polymerization of any polymer is particularly constant and does not depend on the type of initiation. Activation energy for elementary stages at polymerization of olefins for the process of the polymer formation with the particular rate E_R and polymerization degree $E_{\overline{X}}$ is determined as follows:

$$E_R = E_i + E_g + E_b$$
$$E_{\overline{X}n} = E_g - E_b$$

where E_i, E_g, E_b - energies of activation of elementary stages: initiation, growth and break of the chain, respectively E_b is changed by E_{tr}, if the chain break proceeds by the chain transfer. The chain growth at cationic polymerization in the indifferent medium requires no high activation energy. Values of E_i and E_b are principally higher than E_g. The value of E_R for many polymerizing systems is negative (in the range of 20.5 - 42 kJ/mole). However, negative activation energy at cationic polymerization is not a law. The sign and value of E_R varies in dependence on the nature of monomer, catalyst and other corresponding agents. Even for one and the same monomer the value of E_R may change depending on the application of the catalyst, the promoter and the solvent. The differences in E_R are explained by the changes in the values of E_i, E_g and E_b, caused by the differences in the nature of AC, solvating reactivity, etc. The rates of cationic polymerization does not change sufficiently in dependence on the change of the temperature of the process performance. Activation energy E_R with the polymerization order $\overline{X}_n$ is always negative, because $E_b > E_g$ in all the cases, independently on the method of the chain break. In this case $\overline{X}_n$ decreases with the increase of the polymerization temperature.

Various aspects of cationic polymerization of olefins are presented sufficiently full in the numerous papers, in particular in [2, 3, 6, 7, 8]. However, in spite of great number of investigations there is sufficiently small amount of direct proofs of the composition of carbonium ions (R^+), participating in cationic polymerization. This circumstance is explained by high lability of R^+. That is why their identification is sufficiently difficult, because, for example, there often occurs the interaction of carbonium ions with counterions in the system; they are isomerized easily, forming sufficiently stable structures. In this connection and in regard with their instability is often very difficult task of determination of the concentration of active centres (AC) in equilibrium state. The task is aggravated by one more fact, that the reaction extremely sensitive to admixtures must be performed in carefully controlled conditions. Complex Brenstedt acids are sufficiently often the catalysts of cationic polymerization of olefins, are characterized by specific features, which define the mechanism of the process initiation and are not clear enough till the present time, as well as the role of counterion in this process. It would seem that the task of identification of carbonium ions in the reactions of polymerization was simplified after proving that the spectrum f the system 1,1-diphenil-ethylene/(BF$_3$)/H$_2$O) in benzene is similar to the one of the ion of 1,1-diphenilethylene carbonium [7]. The system BF$_3$·H$_2$O is widely used as the initiator of cationic polymerization and that is why the data on simple protonation of olefin with the participation of this system are of great importance for the development of general theory. However, these results are in contradiction with the fact that proton detachment from the aquacomplex BF$_3$·H$_2$O is energetically unprofitable [9]. Consequently, the possibility of usual proton transfer to the molecule of the monomer is excluded. Moreover, there exist no direct spectroscopic proofs of the presence of free protons or carbonium ions in similar polymerization cationic systems. Other methods of identification of carbonium ions are less precisional, circumstantial and mostly often qualitative ones.

Identification of carbonium ions by the method of electrical conductivity does not also give any clear results even because, for example, for the determination of the charge carriers it is necessary to know that such particles are inassociated. Moreover, the part of carbonium ions, presenting in the reactionary system depends on the reaction temperature, dielectic penetrability and the number of other factors (viscosity of the system, chemical nature of the solvent, etc.). It may be expected (and the experiment proved this) the validity of the statement that the conductivity of the system must increase with the growth of the solvent polarity. It is more difficult to explain the fact sometimes observed of the increase of the system at the end of the polymerization process. According to this cause at the study of the complex systems it is necessary to be careful with the complete conclusions.

In 1932 for the first time it was suggested the mechanism initiation of electrophilic reactions by H-acids, the sense of which is reduced to usual, for the initial view, attachment of carbcation to π-bond of olefin:

$$H^+, A^- + CH_2{=}CHR \rightarrow CH_3{-}C^+HR, A^-$$

This is a biomolecular reaction, in which H^+ (or carbcation) attacks mostly hydrogenized atom of carbon of the monomer, i.e. the atom with maximum electron density (according to Markovnikov's rule), giving the most stable thermodynamically carbonium ion. The activity of protons may be sufficiently supressed by the solvation effects. And this, in its turn, allows to explain the dependence of H-acids on the medium polarity and the reaction conditions.

Among protonic acids (Brenstedt acids) the catalysts, based on halogenides of metals MeX$_n$ with proton donors, for example with water, alcohols, alkylhalogenides, are sufficiently effective initiators of cationic polymerization. The activity of such H-complexes is sufficiently stipulated by chemical composition of cocatalyst (H$_2$O, ROH, HCl, HF, etc.) with the participation of which AC is formed. In particular, the mechanism of the initiation reaction at cation polymerization in the presence of the aquacomplex of BF$_3$·H$_2$O may be presented by the following scheme:

$$H_2O + BF_3 \underset{\leftarrow}{\overset{\rightarrow}{}} \Big[\underset{H}{\overset{H}{\diagdown \diagup}} O - BF_3 \Big] \underset{\leftarrow}{\overset{\rightarrow}{}} H^+, BF_3 \bullet OH^-$$

$$+H^+, BF_3 \bullet OH^- \rightarrow \Big[\overset{CH_2}{\big\| \longrightarrow} H^+, BF_3 \bullet OH^- \Big]$$

$$\underset{R}{\overset{\diagup CH_3 \diagdown}{}} \underset{\underset{R \quad H}{\diagup \diagdown}}{C^+} \overset{H}{} \quad BF_3 \bullet OH^-$$

However, in fact this scheme of the first glance simple reaction is out of any critics, because the existence of the system H^+, $BF_3 \cdot OH^-$ seems to be hypothetic, as it does not take into account the active role of the monomer.

If the complexes with elements possessing d-orbitals, for example AlX, instead of boron compositions (see the scheme), then there appears the possibility of additional interaction as a result of overlapping of saturated d-orbitals of acceptor with vacant antibonding orbitals of the olefin. Such interactions stabilize and strengthen H-complexes, increase electron density of the olefin, making difficult the formation of AC. Probably, complex H-acids, forming on the basis of halogenides of metals and proton donors may be subdivided into the ones, containing d-orbital and H-complexes, possessing no d-electrons. One more feature of such catalysts is the possibility of their poisoning by promoters (cocatalysts). The concentration of cocatalyst, at which the efficiency of electrophilic catalyst is maximal, depends on the nature of the promoter, monomer, solvent and the conditions of the reaction performance. The excess of the cocatalyst, which basicity is higher than for the monomer, competes for the complex protonic acid, forming in the equilibrium reaction, for example:

$$H_2O + AlCl_3 \underset{\rightarrow}{\overset{\leftarrow}{}} H^+[AlCl_3 \bullet OH]^- \xrightarrow{\;H_2O\;} [H_3O]^+[AlCl_3 \bullet OH]^-.$$

In the case of application of water as a promoter there is formed the salt of oxonium ion, which activity relating to the monomer is sufficiently lower than the activity of complex protonic acid, there exist no reliable data on numerical values of individual rates of initiation (and not total rates of polymerization). This does not allow to estimate the degree of the validity of the scheme suggested of the process of cationic polymerization. Concentration of the promoter influences also the molecular weight of forming polymer (participates in the reaction of the chain transfer). That is why it is sufficiently probable, that the effect of the catalyst poisoning by the excess of cocatalyst may be connected with the reactions, proceeding after the stage of initiation, and not before it. Immediately when growing carbonium ion is formed in the presence of water excess, for example, there begins the competition between the chain growth with the participation of the monomer and the process of the chain restriction with the participation of cocatalyst [7]:

$$\sim C^+X^- + M \underset{growth}{\overset{chain}{\longrightarrow}} \sim CM^+X^-;$$

$$\sim C^+X^- + H_2O \underset{restriction}{\overset{growth}{\longrightarrow}} \sim C - OH + HX.$$

In the case of realization of the second reaction total rate of the process of polymerization decreases, the formation of the polymer being stopped probably.

The unique feature of the chain growth is the fact, that growing polymeric chain, forming after the attachment of the initial chain of the monomer, is practically completely similar to the polymeric chain, formed at the attachment of the next monomeric chains. In cationic polymerization AC is strong electrophil. Growing carbonium ion R^+ during the process of polymerization is the ion, possessing one p-orbital, formally localized on the end carbon atom, possessing no electrons. In fact, usually sufficient influence on the rate and trajectory of the process is performed by the conjugation effects. In this connection electroshortage of the considered carbon atom decreases a little. For example, at polymerization of alkylvinyl ester positive charge is delocalized on the oxygen atom and on the end carbon atom. In the general case the chain growth is stipulated by a specific combination of electrostatic and orbital interactions in the reaction between carbcation and olefin, which is controlled by counterion.

Contact ionic pairs in equilibrium state, separated by the solvent, and free ions may exist in the polymerizing system in the presence of various solvents. There are no clear proofs of the participation of these AC in concrete process at the polymerization, yet. Evidently, different ionic structures of AC are in dynamical equilibrium, which depends on temperature, concentration of the reagents, the nature of the catalyst and the solvent. In this connection the kinetics of cationic polymerization may be unpredictable and change suddenly at the change of external conditions of the reaction. It is known, that there are possible the changes of stereochemical composition of cationic end of the polymeric chain during its growth at cationic polymerization. This causes the higher increase of the probable amount of various AC. The characteristic example may be shown by the ability of carbonium ions for isomerization. At cationic polymerization of 3-methylbutene-1, for example, in presence of $BH_3 \cdot H_2O$ there proceeds a spontaneous transfer at the temperature change:

$$
\sim CH_2 - \overset{\displaystyle |}{\underset{\displaystyle \underset{/\,\backslash}{\underset{H_3C\quad CH_3}{CH}}}{C^+H}} \underset{\leftarrow}{\rightarrow} \sim CH_2CH_2{}^+C \overset{\displaystyle CH_3}{\underset{\displaystyle CH_3}{<}}
$$

In this case it is said about isomerizational polymerization [5]. Beside different kinetic behavior of these ions, the chain growth on each of them leads to polymers with different structure of macromolecules. If one and the same AC changes its chemical structure several times during the growth reaction, then forming macromolecules will represent the block-copolymer with different stereochemistry in the sequence of monomeric chains in macromolecules.

The presence of different types of AC in the system naturally makes the kinetics of polymerization process difficult, especially in the cases, when the equilibrium between different ions is displaced during the reaction proceeding. Until now there are no methods worked out in cationic polymerization of reliable identification of different types of AC. That is why the values of the rate constants of the chain growth changing are sufficiently approximate, and sometimes wrong.

Apparently, in the most of cationic systems the role of active centres is played by ionic vapors. The feature of the chain growth on such AC is the dependence of the rate constants of the growth (K_g) on the reaction conditions, the nature of the solvent and the catalyst. The differences in K_g for ionic pairs are stipulated by the structure of AC, participating in the process of cationic polymerization. The reaction ability of AC depends on the nature of counterion and on the distance between the components of ionic pair, depending on its turn on the medium polarity, its ability to specific solvatation of the components of ionic pair by the solvent (or by the monomer), and by the reaction temperature also. Quantitative data, reflecting the influence of counterion on the kinetics of the process of cationic polymerization, are unsatisfactory. But their obtaining is made difficult, because usually the concentration of growing ionic pairs is unknown as well as the constant of the catalyst dissociation into free ions. The last circumstance may introduce sufficient mistakes into measured values of K_g because of high reaction ability of free ions, formed from ionic pairs. Activation energy of the chain growth E_a does also sufficiently depend on the dissociation of growing ionic pairs. If negative counterion interacts with the monomer before its reaction with the cation, and this interaction is displayed by the formation of instable intermediate product, then the rate of the chain growth may be defined by stationary concentration of this product and the rate of its formation. Classical cationic reaction of the chain growth is electrophilic, but if its rate is defined by preliminary interaction with negative counterion, the reaction display nucleophilic character.

From the practical point of view the reaction of the chain growth restriction influences negatively the process of the polymer obtaining, limiting both the yield and the molecular weight of forming products. At the chain transfer the number of AC does not change. At the chain break the growth of the initial chain stops and carbcation is destroyed (or stabilized). In this connection reinitiation is impossible. In this case the number of AC decreases and the polymer is formed with lower yield and the molecular weight. Sometimes these reactions are difficult to distinguish.

The cases are possible, when active particles are generated from AC simultaneously with the formation of "dead" polymer. These particles differ from the initial ones by their reaction ability. If the difference in the reaction ability is high the inhibiting of the polymerization reaction becomes very efficient. But if such difference is small, the reaction may be considered as the reaction of the chain transfer. Consequently, the difference between two processes is defined by kinetic factors. Both reactions of the chain restriction may proceed in parallel also.

Thus, even considering the general picture of the polymer formation according to the scheme of cationic polymerization, the difficulties in understanding thin mechanism of separate elementary stages are completely clear. First of all, this relates to the acts of initiation and growth of polymeric chains in low polar hydrocarbonic media in the presence of complex Lewis and Brenstedt acids, reflecting high specificity of the reactions of the AC formation.

CHAPTER 2

CATALYTIC ACTIVITY OF LEWIS AND BRENSTEDT COMPLEX ACIDS AS ACTIVE CENTRES OF CATIONIC POLYMERIZATION OF OLEFINS

2.1. Initiation of cationic polymerization of olefins by complexes of Lewis and Brenstedt acids

Complex catalysts of cationic polymerization of olefins, based on Lewis and Brenstedt acids [2, 5-13] form the complexes possessing high acidic strength in comparison with the initial compounds due to the interaction of the components (Table 2.1). The effect of acidic systems strengthening at complex formation is accompanied by a sufficient increase of their catalytic activity [14]. The proof of high abilities of the complex acids is the data from the paper [3], in which "living" carbonium ions were obtained at 193-233°C from olefins nonpolymerizing in usual conditions under the influence of strong complex acids ("superacids") [15].

Table 2.1.

No.	Protonic acid	$-H_o$	No.	Complex acid	$-H_o$
1	HF	9.7	5	$HF(HCl) \cdot BF_3(AlCl_3)$	16.6
2	H_2SO_4	11.9	6	$HSO_3F \cdot SbF_5$	17.5
3	HSO_3Cl	12.8	7	$HF \cdot 9SbF_5$	20
4	HSO_3F	13.9	8	$HF \cdot SbF_5$	>20

On the other hand, complex formation of individually inactive protogenic compounds, for example alcohols, with corresponding Lewis acids leads to the formation of the catalysts, characterized by aimed performance of cationic polymerization of olefins [17].

Complex catalysts should be subdivided according to their acidic-catalytic properties. It is considered the complexes of strong H-acids and weak proton donors with Lewis acids.

To the complexes of strong H-acids the compositions relate, formed at the interaction of Lewis acids ($AlCl_3$, BF_3, $SnCl_4$, $FeSO_4$) and Brenstedt ones of HX type (X-Cl, F, etc.) [3, 16], characterized by the sufficient acidic strength (Table 2.1). The condition of the existence of complexes is the stabilization of a proton, stipulated by its interaction with the solvent or the base, and the formation of relatively stable counterion [18, 19]. The complexes are formed depending on the nature of Lewis acids and some other factors, which differ by the degree of H-X bond friability. They are placed in the following sequence according to the increase of acidic-catalytic properties:

a) Complexes with different activating solid surfaces [20-23]:

$$HX + Z \rightleftharpoons H^{+\delta} \ldots [X - Z]^{-\delta} \qquad (2.1)$$

where Z is the solid activated surface (silicagel, kyzelgur, aluminum oxide, etc.).

b) Complexes with metal salts (sulphates) [23-25]:

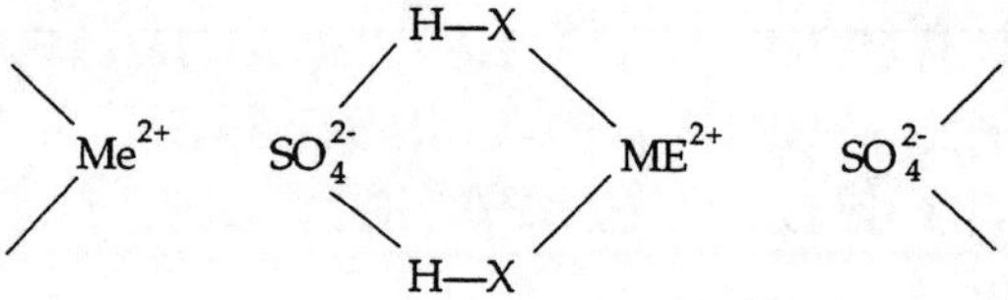

$$(2.2)$$

where Me=Zn, Sn, Pb, Cu, etc.

c) Complexes of proton attachment or Gustavsson complexes (GC) [16, 26-29]:

monomer's

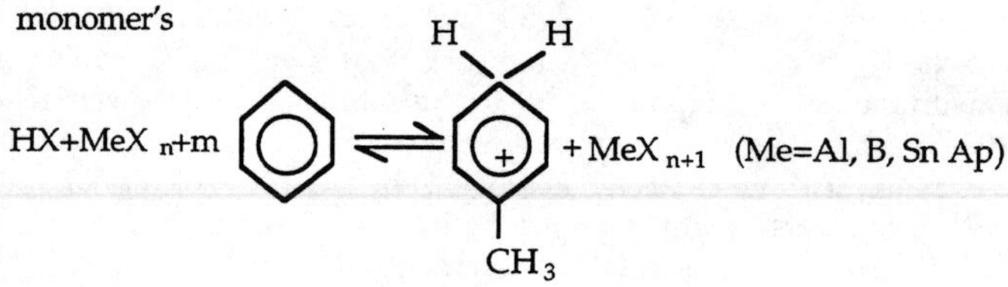

$$HX + MeX_{n+m} \rightleftharpoons \cdots + MeX_{n+1} \quad (Me=Al, B, Sn Ap) \qquad (2.3)$$

polymeric

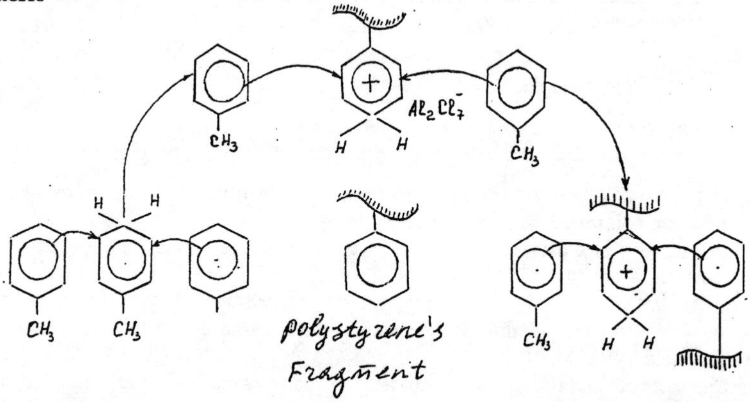

Solid carriers Z play the role of "adsorptional layer" for X ions, promoting the activation of HX molecules (reaction 2.1) [7, 20-22]. As a consequence of HX interaction with Z the concentration of protons increases (the rate of initiation increases) of the polymerization of vinyl esters, and the reaction of X^- with growing polymeric carbcation decelerates. The strength of the acids, attached to silicagel increases in the sequence: $H_3BO_3 < H_3PO_4 < H_2SO_4$ [20]. The attached phosphoric acid is characterized by great amount of acidic centres in the range of $-5.6 < H_0 < -3$, at the same time sulfuric acid being characterized by the range of $-5.6 > H_o > 8.0$. Even the acidic centres of these acids relate to Brenstedt type, nevertheless their activity is defined by the properties of the carrier. The influence of the solid carrier of H_2SO_4 is specially characteristic in the polymerization of alkylvinylic esters, leading to the obtaining of stereo regular polymers [20, 21]. The following characters increase in the sequence

$$CoSO_4 < FeSO_4 < MgSO_4 < Fe_2(SO_4)_3 < Cr_2(SO_4)_3 < Al_2(SO_4)_3$$

with the growth of the interaction between the molecules and the surface of $MeSO_4$: the catalyst acidity, the rate of vinylisobutylene ester polymerization, molecular mass (MM) and stereo regularity [20]. Taking into account low catalytic properties of sulphates and the increase of activity of the catalysts with the decrease of the amount of free acid, the authors explain the results obtained from the point of view of the complex formation of acid anion and metal salt. As a result the surface plays the role of strong protonic acid, and counterion is a weak nucleophil. Another type of activation of H-Cl bond is possible in the complexes of chlorinated hydrogen with metal sulphates, which is concluded in the simultaneous interaction of HCl with cation and anion of the salt (reaction 2.2). In this case complete ionization of H-Cl bond does not occur. The states of HCl in complexes $MeSO_4 \cdot mHCl$ may be presented schematically by one of the following structures [23]:

$$Cu^{2+}\begin{bmatrix} H_2SO_4 \\ 2Cl \end{bmatrix} \quad Cu^{2+}\begin{bmatrix} SO_4^{2-} \\ 2Cl \end{bmatrix} \quad Cu^{2+}\begin{bmatrix} SO_4^{2-} \\ 2HCl \end{bmatrix}$$
$$\text{I} \qquad\qquad \text{II} \qquad\qquad \text{III}$$

The structure III with polarized HCl molecules is the most probable one. In particular, this is testified by the data of IR-spectroscopy. The absorption bends of $CuSO_4 \cdot 2HCl$ at 2615 cm^{-1} and 2500 cm^{-1} are corresponded to valent oscillations of H-Cl and are displaced to long-wave range by 1150 cm^{-1} at deuteration. New absorption bends occur (1235, 1180, 1080, 1015 cm^{-1}) accompanying absorption bends of copper sulphate, that is explained by the interaction of SO_4^{2-} anion with HCl. The presence of hydrogen bond leads to the displacement of the absorption frequencies of HCl itself (in comparison with gas product) as well as to noncoincidence of the absorption bends of SO_4^{2-} in corresponding complex compositions.

Acidic-catalytic properties of the catalysts complexes $MeSO_4 \cdot mHCl$ (Me=Zn, Sn, Pb, Cu, Hg, m=0.2-1.7) depend on electronegativity of the salt cation X_i [25]. Thermal stability of the complexes to HCl elimination increases with the increase of X_i of the metal cation in the sequence of complex catalysts $ZnSO_4 \cdot 0.2HCl$; $SnSO_4 \cdot 1.1HCl$; $PbSO_4 \cdot 1.1HCl$; $CuSO_4 \cdot 1.7HCl$; $HgSO_4 \cdot 1.1HCl$, and the yield and MM of products of isobutylene olygomerization increases simultaneously. The activity of $MeSO_4 \cdot mHCl$ complexes is connected with the formation of metastable structures, being formed during the interaction of HCl with coordinationally nonsaturated metal atoms [25]. Polarization of H-Cl bonds at coordination simplifies the interaction of hydrogen atoms with the bases. In this case the surface of $MeSO_4 \cdot mHCl$ receives negative charge (by means of Cl$^-$ anions) and plays the role of electrostatic stabilizer of carbonium ions in analog to polar solvent. Simultaneous combination in the complexes of the properties of acidic surface and negatively charged polar medium explains the initiating activity of the complexes [25]. The increase of the basicity of the olefin causes the increase of the polarization degree ($\overline{P}_n$) of forming polymeric products in the sequence: isobutylene ($\overline{P}_n$=2-6)<styrene ($\overline{P}_n$=30)<vinyl-butylic ester ($\overline{P}_n$=300-2000). Nonhigh efficiency of the initiation (f=0.2-0.3) points out the fact that only a part of acidic centres is active, which is proportional to the content of HCl in the complex. The yield of the reaction products decreases down to the inhibition of the process in the presence of diethyl alcohol according to the increase of the basicity of the solvent.

The selectivity of the $MeSO_4 \cdot mHCl$ action in the process of selective polymerization of isobutylene is connected with the decreased acidic strength of the complex catalysts (3<H_0<-8), comparing with H_2SO_4(H_0=-11.9).

Ionization of H-X bond takes place in Gustavsson complexes, that defines high acidic strength (H_0 up to -16) (reaction 2.3), [16, 26]. In relative conditions this leads to the formation of the arenonium ion and complex counterion. The energy gain by means of solvatation of the complex by the excess of arene and stabilization of arenonium ion by aromatic hydrocarbons makes the occurrence of the ionic pair easier, that stipulates astechiometric composition of the Gustavsson complexes (it contains 3-4 moles of the aromatic hydrocarbon in the case of low molecular complexes). According to high acidic strength Gustavsson complexes display increased activity in polymerization of olefinic and vinyl monomers in comparison with the above considered HX complexes. This is displayed by the formation of polymeric products with higher MM (isobutylene, styrene, polyvinylbutyl ester) and by the ability to cause the oligomerization of propylene. The effects pointed out are connected with the higher stability of the complex counterion to decay into components under the influence of carbonium ion: $M_n^+ \| AlCl_4 \rightarrow M_n Cl + AlCl_3$, as well as with the acidic strength of GC. The character of ionic part of GC (positive charge delocalized according to aromatic system) explains the specific features of these complexes as electrophilic agents. Thus, proton donor activity of the Gustavsson complexes changes in the sequence: $3C_6H_3CH_3 \cdot HCl \cdot Al_2Cl_6$>4 $(CH_3)_3C_6H_5 \cdot HCl \cdot Al_2Cl_6$>$3(CH_3)_6 \cdot HCl \cdot AlCl_3$.   This sequence is antibathic to the one of the basicity of aromatic hydrocarbons [28]:

$$(CH_3)_6C_6>(CH_3)_3C_6H_3>CH_3C_6H_5.$$

The role of basicity of aromatic component of the Gustavsson complexes at the initiation of cationic polymerization of olefins is not uniform, because its growth causes the increase of stability of arenonium ion and the decrease of the strength of the bond between carbonium ion and counterion [26].

Polymeric Gustavsson complexes are composed similar to their low molecular analogs. However, the presence of aromatic nuclei in their solvate cover, connected with the polymeric chain (polystyrene, styrene copolymer with divinylbenzene), leads to the change of stechiometric composition of the complex, its phase state and some other properties [29]. Initiation of the process of olefins polymerization is reduced to the acception of a proton from areninium ion, showing no dependence on the composition of the Gustavsson complexes. Free aromatic hydrocarbon, being formed from arenonium ions in consequence of H-initiation may be alkylated by growing carbonium ion. The reaction of the polymeric chain restriction by means of aromatic hydrocarbon leads to the formation of new (polymeric) σ-complexes, being more stable than the initial Gustavsson complexes, and being able to reinitate polymerization of olefins in definite conditions:

$$M_n^+ \; + \; \text{(benzene ring, R)} \longrightarrow \text{(complex: } M_n^+, H, +, R) \xrightarrow{M} M^+H \; + \; \text{(benzene ring, } M_n, R)$$

Consequently, the mechanism excitation of cationic polymerization of olefins by the complexes of strong H-acids is connected with competitive interaction of the olefin with counterion (stabilizing bases) in the reaction with counterion, i.e. with the proton transfer from weaker acceptor to stronger one.

On the contrary, such mechanism of excitation of cationic polymerization of olefins by the complexes of Lewis acids with weak proton donors (H_2O, ROH, etc.) is doubtful [9]. Sufficient energy of the break of O-H (O-R) bond of water (alcohol) [30] and, respectively, low acidic strength of the complexes does not allow to explain acidic-catalytic properties of these systems by the Brenstedt activity only.

Other chemical processes proceeding in parallel to the formation of the complexes (alcoholysis, hydrolysis, etc.) and instability of the complexes themselves makes the analysis of their behavior difficult.

Aquacomplexes of four-chlorinated tin relates to the mostly well studied objects of this type [2, 31]. IR-spectrum the polymer possesses absorption bends at 2180 cm^{-1}, relating to oscillations of C-D bond, at polymerization of α-methylstyrene in the presence of $SnCl_4 \cdot D_2O$ complex. The increase of the initial rate of cationic polymerization at the introduction of H_2O points out the initiation of the process by means of the attachment of a proton to the monomer molecule. However, the obviousness of the initiation of the polymerization reaction by a proton is masked by hydrogen chloride, extracted in this case. Th activity of $SnCl_4$ aquacomplexes, nevertheless is connected with the existence of di- and monohydrates of $SnCl_4$, even if they are instable compounds and it was failed to extract them individually [32]. Basing on the analysis of the kinetic data the authors of the work [32] suppose the formation of π-complex between trichlorstannic acid hydrate and a monomer and its further ionization in nonpolar solution (CCl_4):

$$SnCl_3OH \cdot H_2O + C_6H_5CH = CH_2 \rightleftarrows SnCl_3OH \cdot H_2O \cdot C_6H_5CH = CH_2$$

$$SnCl_3OH \cdot H_2O + C_6H_5CH = CH_2 + C_6H_5CH = CH_2 \rightleftarrows HM_2^+ SnCl_3(OH)_2^-$$

$$SnCl_3OH \cdot H_2O \cdot C_6H_5CH = CH_2 + C_6H_5CH = CH_2 \rightleftarrows [HM_2^+ SnCl_3(OH)_2^-] \, solvat.$$
$$C_6H_5CH = CH_2$$

The weak side of the presented scheme is the nature of AC of polymerization. As tristannic acid hydrate is solvated by a solvent molecule at the extraction from ester solutions, coordination number of Sn in $OHSnCl_3 \cdot H_2O$ complex is four, in another case i t must be admitted that $OHSnCl_3$ hydrate is not the real active centre of cationic polymerization of olefin (styrene).

Excitation of isobutylene polymerization under the influence of $SnCl_4 \cdot H_2O$ is connected with the formation of triple complexes $CH_2C(CH_3)_2 \cdot SnCl_4 \cdot H_2O$ [33].

Depending on the water amount in the system the process proceeds with the deceleration of high initial rate (at $(H_2O) \geq (SnCl_4)$) or, in opposite, with the acceleration at the initial moment (low content of H_2O), that is explained by the formation of initiating complexes of two types:

$$\text{SnCl}_4 + \text{H}_2\text{O} \rightleftarrows \text{SnCl}_4\cdot\text{H}_2\text{O} \begin{cases} \xrightarrow{\text{H}_2\text{O}} \text{SnCl}_4\cdot 2\text{H}_2\text{O} & \text{(A)} \\ \xrightarrow{\quad M \quad} \text{SnCl}_4\cdot\text{M}\cdot\text{H}_2\text{O} & \text{(B)} \end{cases}$$

The increase of concentration of the monomer and $SnCl_4$ at the constant concentration of water promotes the proceeding of the reaction to the side of the complex (B) formation. It is accepted that initiating properties of the complex B – $SnCl_4\cdot Mn\cdot H_2O$ are higher in comparison with the complex A – $SnCl_4\cdot 2H_2O$. According to the mentioned it is observed the acceleration of isobutylene copolymerization by styrene, adding higher active isobutylene. This is in relation with the experiments, in which it was discovered the accelerating influence of polystyrene additives, synthesized *in situ*, on styrene polymerization. On the other hand, it is clear the absence of acceleration at polymerization of α-methylstyrene, which forms the complex of B type in the very beginning of the process because of its high activity at the complex formation with Lewis acids. The latter works [34-36] proved the defining role of $SnCl_4\cdot H_2O$-monomer complexes.

Consequently, at the excitation of cationic polymerization by aquacomplexes it is possible the formation of triple $SnCl_4\cdot H_2O$-monomer complexes, which are the most probable AC. However, the kinetic data existing are just circumstantial proofs of the existence of triple complexes and do not disclose the detailed mechanism of cationic polymerization of olefins.

The more problem is in determination of the role of R_nAlCl_{3-n} (n=0-3) complex formation with weak proton donors (H_2O, R'OH) in connection with the excitation of cationic polymerization of olefins. Water and alcohols enter donor-acceptor interaction with aluminum compositions, that is similar to other electron-donor compounds. However, at moderate temperatures the interaction is evidently reduced to nucleophilic assistance from the side of the heteroatom proton donating compound in the reaction of Al-C bond splitting. As a consequence, various hydroxiderivatives, alumoxanes, oxides, etc. may be the produces of the interaction of R_nAlCl_{3-n} with H_2O in dependence on the correlations of components and the reaction conditions [37]. Nevertheless there were disclosed the conditions of the formation and was proved the existence of 1:1 donor-acceptor complexes of H_2O and ROH with R_nAlCl_{3-n} [17, 38].

The appearance of new intensive bends of the charge transfer with λ_{max}=192, 196, 201 and 210 nm in UV-spectra of aquacomplexes, which were not observed in the spectra of the initial compounds, relates to the transfer of nonlinked electrons of oxygen from P_y-orbital of H_2O to vacant Sp^3-orbitals of Al^{3+}. Weak bend λ_{max}=410 nm (lgε=0.4) is observed at the discharge of complexes. This bend is corresponded to prohibited $n \rightarrow \pi^*$ transfer of electrons from $2P_y$-orbitals of oxygen atom of water to vacant $3d^o_{xy,xz,yz}$-orbitals of Al, able to form π-bonds. This leads to the strengthening of the donor-acceptor bond. Disposition and intensity of the bends at 215 nm and in the range of 250-350 nm, related to electron transfers in Al-Cl and Al-C bonds respectively do not differ from the bends of initial R_nAlCl_{3-n} complexes, i.e. it testifies the conservation of these bonds during complex formation.

New bends in IR-spectra of R_nAlCl_{3-n} complexes with H_2O in the range of 450-600 nm are related to the oscillations of donor-acceptor Al-O bond [17]. Together with noticeable displacements of a number of absorption bends frequencies of Al-O bond (alumoxane) at 630 and 635 nm they confirm the conclusion about the formation of donor-acceptor $R_nAlCl_{3-n}·H_2O$ complexes.

For some complexes, for example $C_2H_5Cl_2Al·H_2O$, the constants of instability were determined: 0.450 l/mole – according to Adamovitch's method and 0.535 l/mole – according to the dilution method. The order of the complex dissociation is $\alpha=0.1$, thermal functions are: $-\Delta H=-53.5$ J/mole and S=24 e·s. The totality of the values mentioned allows to correspond $C_2H_5AlCl_2·H_2O$ to donor-acceptor complexes of medium stability, comparable with the stability of some esterates, with $C_2H_5AlCl_2·H_2O$ in particular.

Depending on the nature of aluminum compound and the solvent (alyphatic or aromatic) it was obtained and identified by spectroscopic methods several forms of aquacomplexes: $C_2H_5AlCl_2·H_2O$, $(C_2H_5AlCl)_2O·H_2O$, $(Cl_2Al)_2·H_2O$ [38]. The formation of aquacomplexes of alumoxane is assisted by the application of aromatic or other electrodonating solvents at the synthesis. Alongside with double complexes there are known triple ones of the similar type, for example $(C_2H_5AlCl)_2O·H_2O·C_6H_6$ ($\lambda_{max}=420$ nm, $lg\varepsilon=2.1$).

The aquacomplexes mentioned in inert solvents are characterized by sufficiently high thermal stability (according to hydrolysis of Al-C bond), low specific electric conductivity ($x\sim10^{-10}Ohm·cm^{-1}$) and are not H-acids according to qualitative reaction with xanton [38]. This correlates with low acidic strength expected for aquacomplexes. High chemical lability is characteristic for aquacomplexes, in particular it is observed their decay in the presence of the excess of electron-donor compounds.

Activity of $R_nAlCl_{3-n}·H_2O$ in a number of cationic processes correlates with the abovementioned complexes properties, including their low acidic strength, and differs from the activity of similar systems, formed *in situ* [17, 39]. For example, in the processes of benzene alkylation by propylene high activity of $RnAlCl_{3-n}$ aquacomplexes is combined with the absence of side (secondary) reactions, and in the case of toluene alkylation isomeric composition of monoalkylation products – isopropyl-toluenes is improved simultaneously. Isobutylene is polymerized into high molecular products even in the presence of sufficient (20-30%) amounts of α- and β-butylenes. Selective polymerization of isobutylene in the mixtures with other C_4 olefins under the influence of $(C_2H_5)_3AlCl_3·H_2O$ served as the basis for the development of the industrial processes of the separation of C_4 hydrocarbons fractions. The increase of the product MM is reached and its chemical composition changes (alternating of isobutylene blocks and isoprene units instead of statistic distribution of monomeric units – in the case of $AlCl_3$) at the copolymerization of isobutylene, the variation of the nature of Al compound in the aquacomplex disclosing additional possibility to regulate MM and other characteristics of copolymers, which define exploitational properties. By the way, on this basis in the USSR a new original dissolving method was worked out for the obtaining of butyl caoutchouc in the presence of aquacomplex of sescvialuminum chloride. Brenstedt activity of the complex, determined on the example of diisobutylene – $C_2H_5AlCl_2·D_2O$ model system, is not evident, taking into account its low H-acidity. In this accordance it should be admitted to be low-probable the ionization of the aquacomplex in the presence of the monomer according to the following scheme [40]:

$$C_2H_5AlCl_2+H_2O\rightarrow H^+[C_2H_5AlCl_2]^-.$$

Apparently, cationic activity of R_nAlCl_{3-n} aquacomplex should be connected with the individualizing of the olefin role in the process of the proton transfer, that may be displayed just on the stage of excitation of polymerization. However, experimental proofs of such mechanism do not exist.

The complexity of the problem of the excitation of cationic polymerization of olefins is characteristic for the complexes of Lewis acids with alcohols also. The distinctive feature of the behavior of these systems is the possibility to display proton donating and (or) carbcationic activity by alcohols [41]:

$$RMeCl_n+R'OH \begin{array}{l} \longrightarrow H^+[RMeCl_n\cdot OR]^- \\[2ex] \longrightarrow R^+[RMeCl_n\cdot OH]^- \end{array}$$

[42-49] point out H-activity of alcohols in the composition of $BF_3\cdot R'OH$ complexes ($R'=CH_3$, C_2H_5, n-C_3H_7, iso-C_4H_9), in particular in the reactions of oligomerization and polymerization of propylene, isobutylene, butylene-2, alkylation of phenol by olefins, etc. Kinetic proofs, set for the benefit of H-initiation, for example ten time increase of the rate of styrene polymerization under the influence of $BF_3\cdot C_2H_5OH$ complex in comparison with $BF_3\cdot O(C_2H_5)_2$ [45, 46], must not be considered as sufficiently convincing.

For $AlCl_3\cdot R'OH$ complexes it is also characteristic the proton donating activity. The authors of the paper [50] suggest classical scheme of H-initiation of cationic polymerization of olefins by $AlCl_3\cdot R'OH$ complexes (where $R= CH_3$, C_2H_5, iso-C_3H_7, tert-C_4H_9):

$$ROH+AlCl_3 \rightleftarrows AlCl_3\cdot ROH \rightleftarrows H^+ + AlCl_3\cdot OR^-$$

According to this it is determined the increase of the catalytic activity of $AlCl_3\cdot R'OH$ complexes with the increase of alcohols acidity [51].

Substituting phenol in the complex $SnCl_4\cdot C_6H_5OH$ by C_6H_5OD the rate of C_4H_9 polymerization in CH_2Cl_2 at 195 K decelerates by (1.5 ± 0.5) times. This may be considered as the strong proof of the proceeding of H-initiation of cationic polymerization of isobutylene.

On the other hand, taking into account the analogue of the behavior of Brendstedt and Lewis acids, it should be expected the possibility of the display of carbcationic activity of alcohol complexes with Lewis acids. It is known, that at the interaction of sulfuric acid with an alcohol carbcation is formed [8, p. 16]:

$$(C_6H_5)_3COH+2H_2SO_4 \rightarrow (C_6H_5)_3C^+ + H_3O^+ + 2HSO_4^-$$

This is proved in particular by the presence of the absorption bend at 293 nm. In the case of less nucleophilic alcohols than Ph_3COH or Ph_2CHCH_2OH, the rates of the reactions with the participation of carbonium ions often depend on the functions of alcohol Hammet acidity H_0. Carbcations appear as a result of double stage process [8, p. 53]:

$$R'OH+B^+H \rightleftarrows R^+OH_2+B \text{ (rapidly)}$$

$$R'OH \rightarrow R^+ + H_2O \text{ (slowly)}$$

Solutions of the third and secondary alcohols in 'super' acids do also form corresponding carbonium ions easily [3, p. 485]:

$$R_3COH \xrightarrow{\quad SbF_5 \quad} R_3C^+SbF_5OH.$$

The determinant for adamantoles in the formation of adamantyl cations under the influence of $BF_3 \cdot COOH$ is the steric effects [52].

There exists the list of experimental works, testifying carbcationic activity of BF_3 complexes with different alcohols in the reactions of alkylation of aromatic hydrocarbons by olefins [53].

The similar result was obtained for the complexes of alcohols with alkylaluminum chlorides also, in particular for $C_2H_5AlCl_2 \cdot ROH$ [41]. In the sequence of alcohols: CH_3OH, tert-C_4H_9, C_6H_5OH, the probability of displaying carbcationic activity increases. Alkylation of benzene by $AlCl_3 \cdot OHCH_3$ complexes proceeds through the intermediate formation of carbcations, that allows to consider this reaction as the model one in relation to active centres of cationic polymerization of olefins [3, p. 259].

Thus, the application of alcohols in combination with Lewis acids discloses the possibilities of the change of initiating properties of catalytically active complexes in the processes of cationic polymerization by means of the change of alcohol radical nature. In reality the analysis of the literature on this question shows the absence of clear experimental data according to this point.

2.2. Characterization of Active Centres of Complex Acids of Lewis and Brenstedt in the Processes of Cationic Polymerization of Olefins.

The analysis of the initiation process of cationic polymerization of olefins is sufficiently difficult task especially if the behavior of the complex catalysts is considered in both chemical and physico-chemical aspects, that supposes the study of the equilibrium of the complex formation of catalyst components in nonwater media. However, physico-chemistry of equilibrium reactions in nonwater solutions and that of acidic-basical interactions, in particular, is difficult to be studied because of the excitation of cationic polymerization of olefins in the media with low values of dielectric penetrability ($\varepsilon=2\text{-}10$). That is why the initiation in cationic polymerization of olefins represents one of the most complex elementary stages in the chemistry of high molecular compounds [54].

The most number of complex catalysts in nonwater media are compounds, low dissociated into ions. They are often associated into dimeric or more complex compositions [54]. From this it follows the indeterminacy of different ionic forms of the catalyst in relation with the ability to the interaction with the monomer in the composition, activity, etc., i.e. multiformity of AC takes place.

According to accepted classification AC in ionic polymerization are subdivided into free and linked ions [18]. Cations of the following types are considered as free ones: H^+, R^+, etc. Linked ions represent ionic pairs, polarized molecules, and nonpolarized intermediate states of molecules:

$$RY; \qquad R^+Y^-; \qquad R^+ | \, | Y^-; \qquad R^+ + Y^-$$
$$I \qquad\quad II \qquad\quad III \qquad\quad IV$$

I - compounds with covalent bond; II - contacting ionic pairs; III - solvate-separated ionic pairs; IV - free ions.

AC of I and IV types are not characteristic for cationic polymerization of olefins [18]. Ionic pairs with different degree of separation are the most wide-spread AC of cationic polymerization of olefins. The contacts of dissociation been measured ($K_D=10^{-4}$-10^{-5}) for some triphenilmethylic salts in media usual for cationic polymerization ($\varepsilon=10\text{-}17$) point out the fact, that the equilibrium of the dissociation reaction is displaced to the side of ionic pairs. On the other hand, it is pointed out the important role of free AC [10]. To determine some characteristics of ionic pairs in nonwater solutions K_D was calculated theoretically, using the known equation of Byerrum-Fuoss [56]:

$$lgK_D=31g(a)-A'-B/\varepsilon Ta.$$

Here a is the distance between ions; ε is dielectric constant of the solvent; T - temperature; A and B are positive constants.

In the initial approximation a does not depend on T and ε, considering $lg(a)<0$. Then:

$$-lgK_D=A+B/\varepsilon Ta.$$

The equation reflects the fact, that the value of anion radius K_D decreases with the decrease of ε. I.e. this allows to estimate the dependence of K_D on the distance between ions, solvent nature and temperature, and on the ratio of different AC also. Ionic pairs may exist in cationic systems in equilibrium with free ions. Despite the fact, that cations may be free from electrostatic influence of anion, their behavior changes in other interactions, such as association with polar or polarizing solvents or monomer [56]. That is why reactivity of such free ions depends on the surrounding and different solvatation degree, changing with the variation of the surrounding geometry and electron density of the reactionary centre. At coexistence of AC as free cations and ionic pairs total reaction rate of polymerization of olefins is the sum of the rates of free cations and ionic pairs themselves [56]. It is necessary to point out, that beside chemical nature of the solvent, temperature and interionic distance the following parameters influence the behavior of ionic pairs also: the size, the charge and the structure of cation and anion. Concurrent with electrostatic interaction, which grows with the decrease of the cation size, solvatation ability of the solvent also influences sufficiently the correlation between the amounts of contacting and solvate-separated pairs [57]. This is displayed by the fact, that for many salts of alkaline metals the part of solvate-separated pairs decreases in the sequence of cations from Cs^+ to Li^+ according to the decrease of solvatation. The size of anion influences sufficiently the dissociation constant, which is higher, in particular for SbF_5 comparing with BF_4 [57]. Differing from anionic AC, sizes of counterions of which are small usually and dissociation becomes more simple with the decrease of the counterion radius, the solvatation of cationic AC is difficult because of big sizes of counterions. The type of substitutors of cations should be corresponded to the structural factors, being able to influence the character of the ionic pairs [58].

At present time there are practically no direct experimental data, allowing to make a conclusion about the structure of various types of AC of complex catalysts. To tell the truth, a definite information about AC may be obtained by circumstantial experimental methods, in particular, from the analysis of the end groups of macromolecules, the spectrum of the products of complex acids transformations, etc. [8, p. 15]. However, circumstantial methods of the investigation do not allow to conclude about the formation processes and AC structure with the sufficient assurance.

The experiment also gives no answer to the question about the content of AC, their strength for homogeneous complex catalysts of polymerization of olefins. The data are practically absent also on the strength of acids in the solvents with low dielectric penetrability [54].

The experimental study of the structures of various types of AC, containing a monomer, was not performed evidently because of their sufficiently high reactivity.

Consequently, the modern level of the knowledges possesses no full information on the connection of the nature of complex catalysts and potential AC. That is why the interest to theoretical study of the behavior of complex cationic catalysts, by semiempyric quantum-chemical ones first of all, is natural.

CHAPTER 3

QUANTUM-CHEMICAL METHODS OF CALCULATIONS OF THE COMPLEXES AS INITIATORS OF CATIONIC POLYMERIZATION OF OLEFINS

3.1. Characteristics of Semiempyric Quantum-Chemical Methods of Calculation

The methods of molecular orbitals (MO) in LCAO approximation (linear combination of atomic orbitals) are the wide-spread theoretical quantum-chemical methodics of the investigation of molecular systems. The great amount of classical works is devoted to general description and the analysis of MOLCAO methods [59-71] for example. All the methods of LCAO approximation are subdivided into two big groups: strict nonempyric (ab initio) and semiempyric SCF methods (of self-consistent field).

Nonempyric calculations, which realization requires the value of fundamental physical constants only, is applied in the present time for three main purposes: first of all, to discover how good the facts set are described by the theory, i.e. to find the limits of the theory application; secondly, for the calculation of the properties of molecular structures or effects, which are difficult or impossible (transient states of the reaction) to determine experimentally; thirdly, to substantiate and developed various semiempyric schemes of calculation. The first two aims are characteristic for semiempyric SCF methods also, in which the experimental data are applied for the estimation of some integrals. Morever, semiempyric methods assume a number of approximations. For example:

1. Some groups of electrons are not considered directly, in particular, only valent electrons are calculated (valent approximations);

2. Some integrals are not calculated or are taken, equaling zero. Sometimes these integrals are expressed through empyric parameters.

However, the mentioned approximations or substitutions may not be of random character. The main criterion of the approximations must be the constancy (invariance) of the calculation at the rotation of coordination axises and translocations (translocations) of the molecular systems from one point of the space to another without structure change. In this connection it is possible the number of circumstantial approximations [66]: a) leading to the fact, that the results of calculations become noninvariant according to rotations of the coordination axises, or to hybridization of atomic orbitals (AO); b) leading to the preservation of invariance of the results according to rotation of the coordination axises, but possessing the break of invariance according to AO; c) preserving the results of invariance according to rotation of the coordination axises and AO hybridization; d) preserving invariance of the results at any orthogonal transformations of the AO basis, including the transition to the symmetry orbitals.

In practice it is sufficiently enough usually the preservation of the invariance of the second and the third level.

All semiempyric methods of calculation may be subdivided into the methods of Complete Differential Overlapping (CDO) and Zero Differential Overlapping (ZDO).

The most wide-spread method from the first group is the broadened Hukkel method (RMX and ENT) [66, 55], in this case it should be taken into account, that this very method does not calculate electron-electron interaction. Comparing with nonempyric methods ZDO approximation possesses the sufficiently less amount of calculated integrals of intermolecular interaction. On the whole, the known variants of ZDO satisfy Pople classification, according to which it may be separated three levels of approximation, preserving sufficient invariance of the solutions for the concrete practical application [66]. The idea is based on the fact, that for all nonempyric calculations double electron molecular integrals of (ab/cd) type, for which the condition a=b and c=d is fulfilled in AO, are always many times higher than the integrals, for which even a single condition is not fulfilled. It was suggested to neglect these integrals. If there is no exclusion for the integral neglecting, the method is called the Complete Neglecting of Differential Overlapping (CNDO). If there are not neglected these of them, which possess orbitals a and b (or c and d) even different, belonging to the one and the same atom, the method is called the Neglecting of Double atom Differential Overlapping (NDDO). If only a part of double atomic differential overlappings is preserved, then we are dealing with Incomplete Neglecting of Differential Overlapping - (INDO) [55].

The methods CNDO and INDO reproduces badly the heats of atomization and orbital energies. Dewar has made a conclusion about the necessity of reparametrization of INDO method in order to obtain reliable values of the heats of the formation and molecular geometry, that promoted the workout of the MINDO method.

The parameters of MINDO method differ from the INDO scheme by the introduction of a number of empyric dependences for the integrals ψ_{AB}, $\beta_{\mu o}$ and the energy of carcasses repulsion, being selected in order to obtain the best correlation between calculated and experimental values of the heats of formation and geometric characteristics of the wide class of standard compositions [10, 11]. Three exist various parametrizations of the MINDO method. The mostly wide-spread is the parametrization MINDO/3, which scheme foresees the optimization of the molecule geometry with the help of gradual lowering to the energetic minimum. In this connection the method MINDO/3 is still the most favourable semiempyric method for building up the surfaces of potential energy and the investigation of the reaction mechanisms. That is why, for example, for completely full study of the complexes of Lewis and Brenstedt acids as the initiators of cationic polymerization of olefins, for the study of the reaction mechanisms and the estimation of acidic-catalytic properties of the complex catalysts it is expedient to use two methods: the MINDO/3 method – parameterizing property – the potential of atom-atomic interaction, and the CNDO/2 – parameterizing property – electron density, q_{H^+} in particular, correlating with the universal index of the acidity of the chemical compound – pKa [4].

3.1.1. CNDO/2 method.

In spite of the great development of nonempyric quantum-chemical methods of the calculations semiempyric classical methods [59-71] do not lose their efficiency until now, and in some cases are well-competing with them [64].

Standard variant CNDO/2 predicts stable geometry of the molecules with the mistake of 0.1 Å and 10° degree. It allows to estimate the charge distributions, differing from the experimental ones by not higher than 0.02 of the electron charge [55], close to the results of nonempyric calculations. Potential curves obtained according to this method are close to the ones obtained from nonempyric calculations [55]. However, it should be taken into account that if the difference of the conformation energies and barriers between them do not exceed several kcal per mole, and in the presence of conjugation in the molecules also, it is possible qualitatively wrong prediction of the geometry. Let us also mention, that the method is inclined to overestimate the stability of twisted conformations also [55].

On the other hand, CNDO/2 method is suitable in approximately equal degree for the estimation of some physico-chemical properties of molecules. Relative simplicity and universality accompanied by predictability, enough for rough estimations, provided the sufficiently wide spreading of the CNDO/2 method. The method allows to estimate the reliability of the results, obtained with the help of it, in particular, comparing them with the data from literature, devoted to similar objects. CNDO/2 method was also applied for obtaining calculations of the conformations and rotational isomerization of the molecules. It was also applied for the calculation of the potential surfaces of many chemical reactions [55].

For better calculation of the donor-acceptor properties single centered integrals (U_{kk}) in the method CNDO/2 are estimated according to the formula:

$$U_{kk} = -\frac{1}{2}(J_k + A_k) - (Z - \frac{1}{2})\gamma_{AA} \qquad (3.1)$$

where J_k - ionization potential;

A_k - electron affinity;

Z - the charge of the atom carcass;

γ_{AA} - integrals of electron repulsion.

Total energy of the system is calculated according to the formula:

$$E_o = \sum \varepsilon_A - \sum \varepsilon_{AB} \qquad (3.2)$$

where

$$\varepsilon_A = \sum^A P_{kk}\left[-\frac{1}{2}(J_k + A_k) - (Z' - \frac{1}{2})\gamma_{AA}\right] + \frac{1}{2}\cdot\sum^A_k\sum_l (P_{kk}P_{ll} - \frac{1}{2}P_{kl}^2)\gamma_{AA}$$

$$\varepsilon_{AB} = \sum^A_k\sum^B_l\left[P_{kl}(\beta_A + \beta_B)S_{kl} - \frac{1}{2}P_{kl}^2\gamma_{AB}\right] + Z'_A Z'_B R_{AB}^{-1} -$$

$$-P_A Z'_B \gamma_{AB} - P_B Z'_A \gamma_{AB} + P_A P_B \gamma_{AB}$$

Here Z'_A, Z'_B - carcass charge of A and B atoms;

γ_{AA}, γ_{AB} - the integrals of electron repulsion;

P_{kl}, P_{kk} - the matrix of electron density;
S_{kl} - the matrix of overlapping integrals;
β - resonance integrals;
R_{AB} - internuclear distance;
P_A, P_B - electron completeness of A and B atoms.

As the values of the total energy of the system - E_o, and the bond energy – E_b, are 2-4 times higher than the real ones, they are used for qualitative estimations only. The bond orders – P_{AB} are determined according to Armstrong [72] with the accuracy of ± 0.02:

$$P_{AB} = \sum_{k}^{A} \sum_{l}^{B} P_{kl}^2.$$

The initial models of complex Lewis and Brenstedt acids are characterized by the number of atoms, the amount of orbitals, total charge of the system (-1, 0, +1) and multiplicity M=2S+1 (S – total spin of electrons).

Minimization of the total energy of the complexes by geometric parameters is performed according to Hook-Jives method and (or) by the method of variable metric [55].

In Hook-Jives method each parameter on each separate iteration obtains some limited increment, and if the function value decreases in this case, new value is used instead of the old one. In opposite case the parameter obtains the increment with the opposite sign. Then the parameter obtains new value of preserve the old one in dependence on the fact if the function decreases or increases. After the sort out of all the parameters it is taken the next iteration and until the moment, when the sort out of the parameters leads to the preservation of its old values.

Optimization would be finished successfully after a single iteration, if the potential pit is stretched along the coordinate axis. In mathematical formulation it means that the matrix of the second derivatives of the energy by geometrical parameters should be diagonal. This may be reached the parameters, describing the molecule geometry, are not bond lengths and angles, but some linear combinations of these values, been named as oscillational coordinates. The lengths and the angles of bonds enter these coordinates in the same proportion, in which they are changing at free oscillation of the molecule. Oscillational coordinates are previously known, however we can imagine the matrix of the second derivatives of functions of L variables by calculating this function approximately in L^2 points, close to the initial one. Possessing such information, it is possible to transfer to new coordinates close to the oscillational ones. In this system coordination relaxation leads to the minimum [55]. This method is the method of the variable metric. It is sufficiently effective especially if the initial conformation of the molecule was obtained wrongly. In this case the procedure of the method of variable metric should be repeated for several times.

To accelerate the process of calculation it is often used the wide-spread method of partial fragmentary optimization [73]. This method supposes that separate bonds, angles and fragments of the models of complexes are set the same as in the known quantum-chemical methods of the calculations of the analog molecules, or they are taken from the known literature sources, formed angles and lengths of the bonds being optimized only.

Sometimes the modified gradient method of optimization is used. The initial point $\bar{x}_o$ is set in multidimensional space of the coordinates R, θ, ϕ of several atoms in the molecule, and the energy of this state is $E(\bar{x}_0)$. It is determined only relative gradients of energy from coordinates (Figure 3.1):

$$gradE_i = \frac{E(\bar{x}_0) - E_i z(\bar{x}, x_i + \Delta x_i)}{|\Delta x_i|}$$

where the increment of the coordinate is calculated as $\Delta x_i = aK_i$, where a is the step of the gradient calculation, K_i – scale multiplicand.

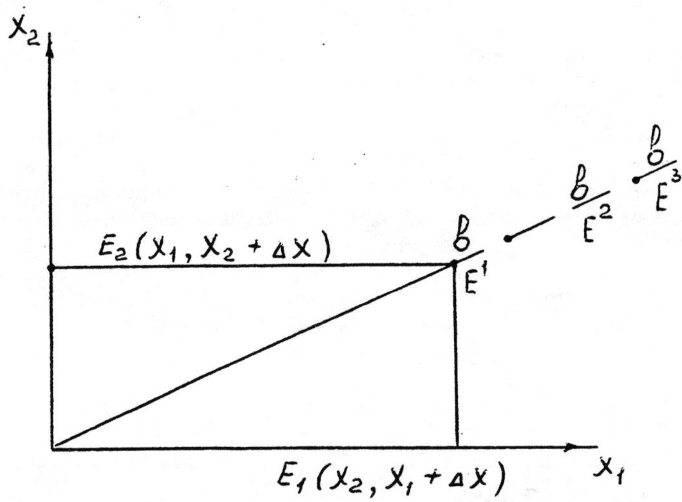

Figure 3.1. Determination of relative gradients of the system energy.

If $E_i(\bar{x}, x_i + \Delta x_i) > E(x_0)$, then $K_i = -K_i$ and again $E_i(\bar{x}, x_i + \Delta x_i)$ is calculated. If again $E_i(\bar{x}, x_i + \Delta x_i) > E(\bar{x}_0)$, then further on this parameter (i) is not varied. After calculating gradients by all coordinates a multidimensional vector is formed according to the value of $gradE_i/\max(gradE_i)$. Then it is moved in the direction of this vector (by straight line with b step, starting from the point E^1) until $E^{\gamma+1} \leq E^\gamma$ at the approximation. The value of E^γ is taken as the new point, and all is started from the beginning (next iteration). Optimization is finished in the case, if: 1) gradients are negative for all parameters; 2) the change of the energy on one iteration is smaller, than the accuracy of the calculation (in order to prevent "skidding" on the plato); 3) no one of the gradients exceeds the minimum energy gradient (also in order to prevent "skipping" on the plato).

The optimizer is intended for the application near the minimum, but it performs badly, if the initial point is set too far from the minimum. Moreover, if the multidimensional surface possesses curved gullying, the wrong point of the minimum may be obtained. In this case it is necessary to make it clear if the calculation is real by means of the repeated calculation, taking for the initial point the one, obtained during the optimization.

To build up the initial model of the molecule of the chemical compound it is necessary to set the coordinates of the corresponding atoms. As it was mentioned above, it is comfortable to determine the coordinates of atoms in so-called molecular coordinate system. To determine the disposition of the atom (as a point) in the space three coordinates are necessary. For example, in Cartesian coordinates they are X, Y, Z. Molecular system coordinates (MSC) allows to operate by the bond length – R, valent angle – θ and by the angle of internal rotation – φ. Three reference points not disposed on one line are necessary at the determination of coordinates in the MSC system. For example, let there be three reference points on the planes: 1, 2, 3 (Figure 3.2). It is necessary, for example, to determine MSC for the atom 4, i.e. values of R, θ, φ. The value of R is determined as the distance from the reference point (atom 1) to atom 4, R being changed from 0 to ∞ and not being negative. Coordinate θ is determined as the angle between vectors 1-2 and 1-4. The angle changes from 0 to 180° according to the definition and may not be negative. Coordinate φ is determined as the rotation angle among the axis 2-1, thus it is taken, that in trans-position (point 4 (0) lies in the list plane) φ=0, and in cys-position φ=180°. In this case positive value of the angle φ is calculated in the direction counter clockwise, if take a glance from the reference point 2 to the reference point 1 (Figure 3.2, φ>0). The range of φ variation changes from 0 to +180° and from 0 to -180°, at φ=+180° and at φ=-180° one and the same point 4 being obtained (Figure 3.2). To provide the MSC calculation for the first three atoms six reference points are set strictly. In order to differ them from atoms they are marked by negative numbers on the selected cube, which edges equal 1 Å (Figure 3.3).

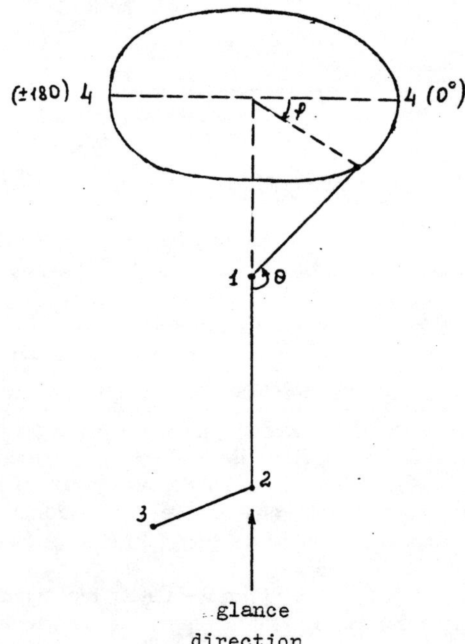

Figure 3.2. Molecular system of coordinates (MSC) (CNDO/2).

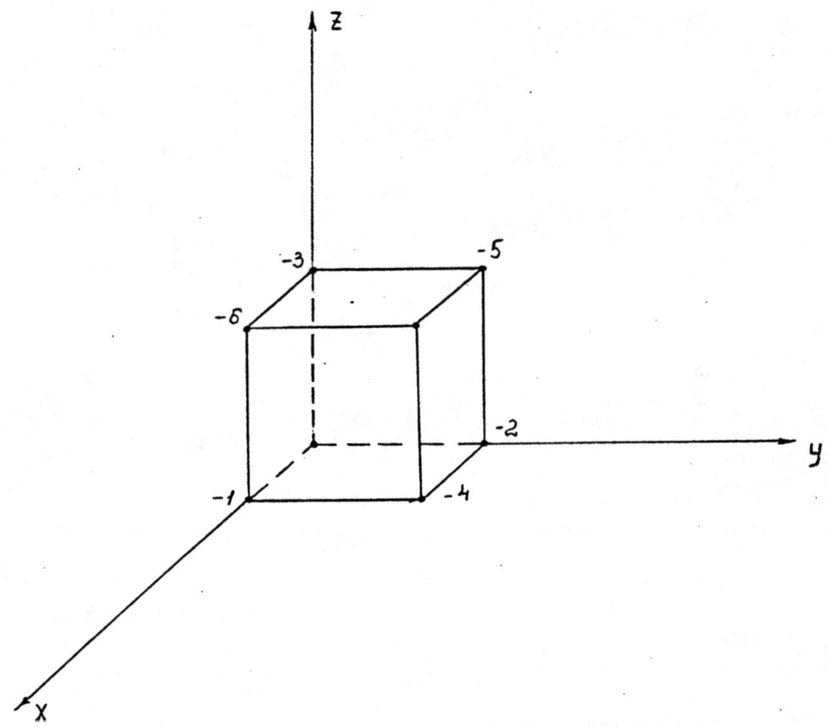

Figure 3.3. The cub of strictly set reference points (CNDO/2).

The rules of determination of the MSC in a molecule (the rules of the tying): a) it is impossible to tie an atom to the atom not tied yet, i.e. it is impossible to select an atom with unknown MSC as a reference point; b) it is impossible to select as a reference points the ones, lying on the one straight line.

Let consider the determination of MSC on the simple ethylene molecule as an example.

Molecule C_2H_4 lies in the plane XOY and one of C atoms is disposed in the beginning of the Cartesian coordinate system (Figure 3.4, a).

To tie C_1 in MSC coordinates it is enough to set three reference points from the six above mentioned. Selecting -1, -4, -2, we obtain the coordinates of ethylene molecule in MSC.

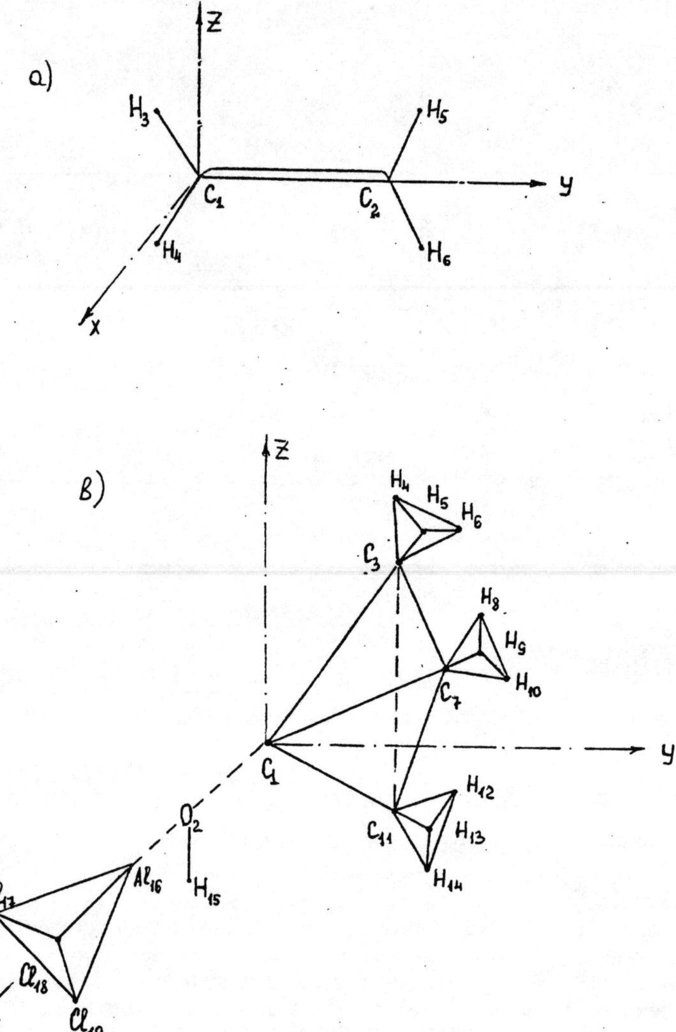

Figure 3.4. Coordinates of C_2H_4 and $AlCl_3 \cdot OH(CH_3)_3$ molecules in the MSC system (CNDO/2).

Table 3.1. Coordinates of C_2H_4 in the molecular system coordinates (CNDO/2).

Atoms	No.*	1 point'	2 point'	3 point'	R(Å)	$\theta°$	$\varphi°$
C1	6	-1	-4	-2	1.00	90.0	180
C2	6	1	-1	-4	1.34	90.0	180
H3	1	1	2	-4	1.12	109.5	0
H4	1	1	2	-4	1.12	109.5	180
H5	1	2	1	-1	1.12	109.5	0
H6	1	2	1	-1	1.12	109.5	180

Note: * – Atomic number according to the Mendeleev's law;
   ' – reference point.

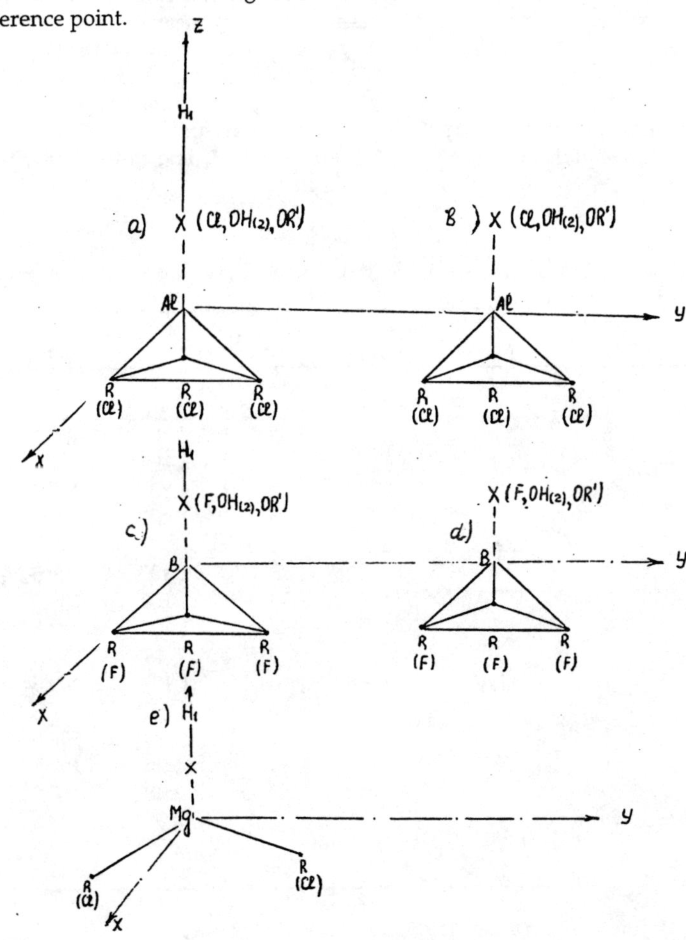

Figure 3.5. Models of complexes, formed on the basis of compounds of Al, B and Mg with some proton-donors.

In more complicated case – $AlCl_3 \cdot OHC(CH_3)_3$ complex, the corresponding atoms are dislocated not on the plane, but in the three-dimensional space (Figure 3.4, b). The corresponding numerical values of the molecular coordinates of the complex $AlCl_3 \cdot OHC(CH_3)_3$ are shown in the Table 3.2.

For theoretical quantum-chemical calculations of the Lewis and Brenstedt acids the following models, based on nontransitional halogenides of metals – $R_nMeHl_{3-n} \cdot HX$ (where Me=Al, B, Mg; Hl=Cl, F; R=OH, CH_3, C_2H_5, i-C_3H_7, t-C_4H_9, n=0÷3; X=Cl, OH, OR) (Figure 3.5).

For the complexes of $R_nAlCl_{3-n}(R_nBF_{3-n})HCl(H_2O,$ R'OH, HF) type the structure may be accepted as tetrahedron or defected tetrahedron, corresponded to the minimum of the energy of the basic state of the complexes. The values of the initial bond lengths and the angles between them should be taken from the literature, for example [21, 22]. Optimization of the complexes structure is performed by means of variation of the bond lengths and the valent angles, that was discussed for many times in the above mentioned (during the consideration of Hook-Jives method, gradient method or the method of varying metric). The selection of the method is defined by complexity of the object for modeling, first of all by the amount of orbitals and atoms. The models without a proton possess total charge of the complexes equal -1, and multiplicity is preserved equal +1.

Table 3.2. Coordinates of $AlCl_3 \cdot OHC(CH_3)_3$ complex in molecular system of coordinates (CNDO/2).

Atoms	No.*	1' point	2' point	3' point	R(Å)	θ°	φ°
C	6	-1	-4	-2	1.00	90.0	180
0	8	1	-2	-4	1.41	90.0	180
C	6	1	2	-3	1.48	109.5	180
H	1	3	1	2	1.107	109.5	0
H	1	3	1	2	1.107	109.5	120
H	1	3	1	2	1.107	109.5	-120
C	6	1	2	-3	1.48	109.5	60
H	1	7	1	2	1.107	109.5	0
H	1	7	1	2	1.107	109.5	120
H	1	7	1	2	1.107	109.5	-120
C	6	1	2	-3	1.48	109.5	-60
H	1	11	1	2	1.107	109.5	0
H	1	11	1	2	1.107	109.5	120
H	1	11	1	2	1.107	109.5	-120
H	1	2	1	-3	1.05	105.5	0
Al	13	2	1	-3	2.20	180.0	0
Cl	17	16	1	-3	2.05	109.5	0
Cl	17	16	1	-3	2.05	109.5	120
Cl	17	16	1	-3	2.05	109.5	-120

Note: * – Atomic number according to the Mendeleev's law;
   ' – reference point number.

It is comfortable to perform the estimation of the acidic strength of complexes according to the values of ΔE_{theor}^{H+} – the energy of proton detachment from the corresponding complex ($\Delta E_{det\,ach}^{H^+}$), q_{H_+} the charge on the hydrogen atoms of H_2O in the complex and δpKa – universal acidity index. It is comfortable to determine the required values of ΔE_{theor}^{H+} ($\Delta E_{det\,ach}^{H^+}$) for the complexes, based on Fridail-Kraphtz catalysts and proton-donors, according to the formulae:

$$\Delta E_{theor}^{H+} \; (kJ/mole) = E_0(R_nAlCl_{3-n} \cdot H_2O) - E_0(R_nAlCl_{3-n} \cdot OH) \qquad (1)$$

$$\Delta E_{theor}^{H+} \; (kJ/mole) = E_0(R_nAlCl_{3-n} \cdot R'OH) - E_0(R_nAlCl_{3-n} \cdot OR') \qquad (2)$$

$$\Delta E_{theor}^{H+} \; (kJ/mole) = E_0(R_nAlCl_{3-n} \cdot HCl) - E_0(R_nAlCl_{3-n} \cdot Cl) \qquad (3)$$

$$\Delta E_{theor}^{H+} \; (kJ/mole) = E_0(R_nBF_{3-n} \cdot H_2O) - E_0(R_nBF_{3-n} \cdot OH) \qquad (4)$$

$$\Delta E_{theor}^{H+} \; (kJ/mole) = E_0(R_nBF_{3-n} \cdot R'OH) - E_0(R_nBF_{3-n} \cdot OR') \qquad (5)$$

$$\Delta E_{theor}^{H+} \; (kJ/mole) = E_0(R_nBF_{3-n} \cdot HF) - E_0(R_nBF_{3-n} \cdot F) \qquad (6)$$

$$\Delta E_{theor}^{H+} \; (kJ/mole) = E_0(R_nMgCl_{2-n} \cdot H_2O) - E_0(R_nMgCl_{2-n} \cdot OH) \qquad (7)$$

$$\Delta E_{theor}^{H+} \; (kJ/mole) = E_0(R_nMgCl_{2-n} \cdot R'OH) - E_0(R_nMgCl_{2-n} \cdot OR) \qquad (8)$$

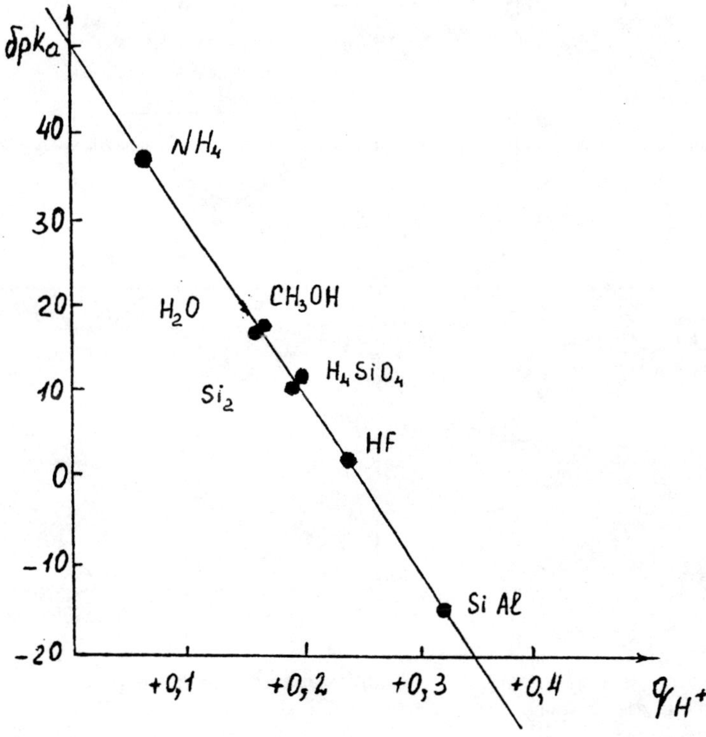

Figure 3.6. The dependence of universal index of acidity pKa on charge on the hydrogen atom – q_{H+} of H-acids (4) (CNDO/2).

The values of δpKa of aquacomplexes are calculated according to correlational ratio [74]:

$$\delta pKa = \beta \cdot \Delta E_{theor}^{H^+} \tag{9}$$

It is admitted that the ratio (9) is correct for both water molecule and for the molecules of corresponding aquacomplexes. Coefficient β equals 0.018 mole/kJ for H_2O ($\delta pKa \sim 15.7$, $\Delta E_{theor}^{H^+} \sim -860$ kJ/mole). Moreover, to analyze the acidic strength of the investigated complexes the correlational dependence between δpKa and q_{H^+} [4] may be applied (Figure 3.6).

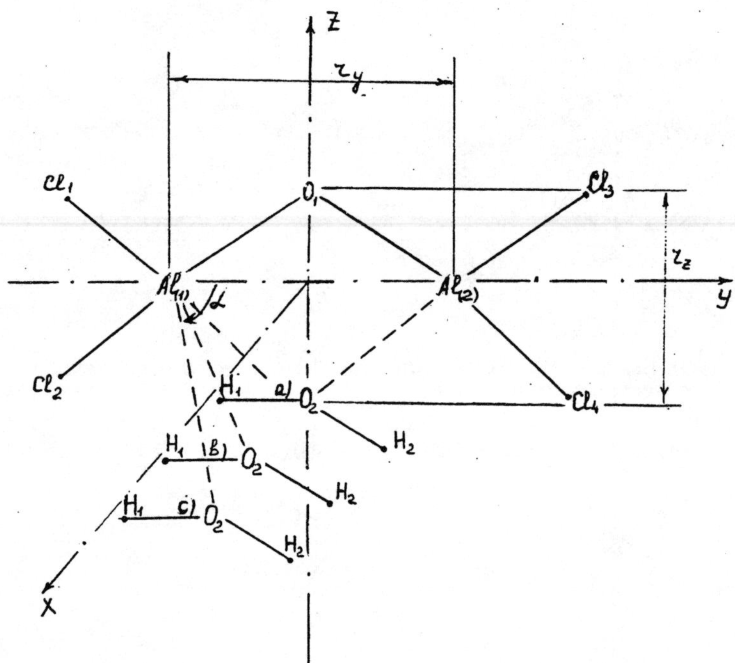

Figure 3.7. The initial model of aquacomplex of alumoxane; a, b, c – dislocation of water molecule in the complex.

The initial models of $R_nAlCl_{3-n} \cdot H_2O(R_nBF_{3-n} \cdot H_2O)$ complexes and alumoxane (boronoxane) aquacomplexes are shown on the Figures 3.7 and 3.8.

The disposition of water in the aquacomplex with alumoxane (boronxane) is defined by coordination of the oxygen atom $O_{(2)}$ in relation to atoms of aluminum (boron). In the initial model of the aquacomplex of alumoxane (boronxane) $Al_{(1,2)}$

$(B_{(1,2)})$ atoms possess S_P^2-hybridization. This structure is close to the one of associated alumorganic (boronorganic compounds), but with different oxygen atoms. It is known for example, that alumoxane possesses bridge-like structure, and the suppression of the bridge is performed by $Al_{(1)}$-$Al_{(2)}$ axis [55]. That is why it is necessary first of all to optimize parameters r_y and r_z, accepting that the bridge is the equal-armed one:

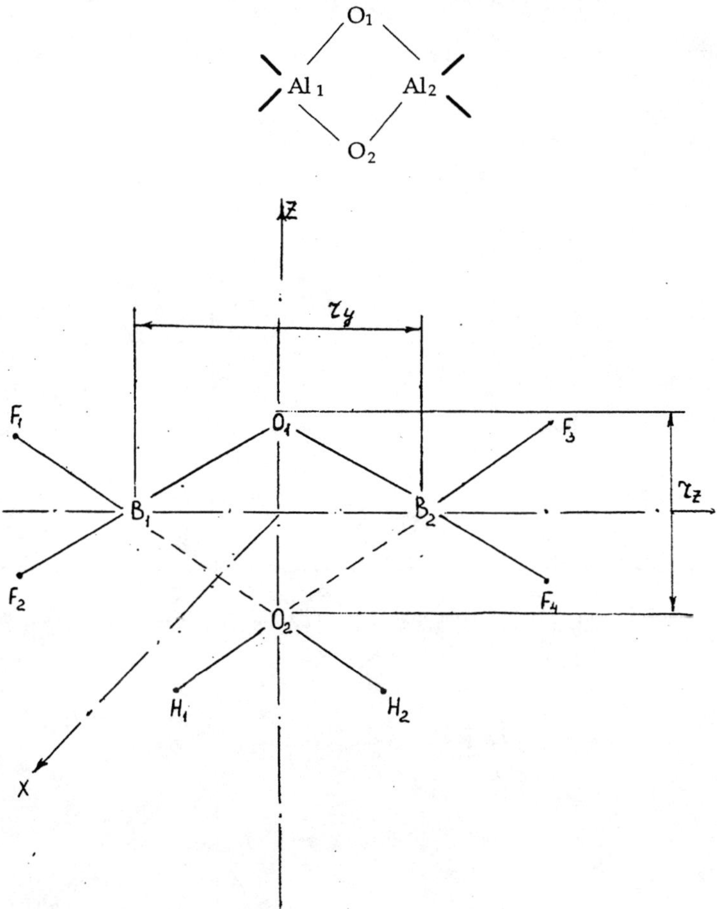

Figure 3.8. The initial model of boronoxane aquacomplex.

Then it is changed the disposition of the water molecule in relation to the $Al_{(1,2)}$ atoms by rotation of H_2O in the plane ZOY by α angle, that relates to the bridge break by $Al_{(2)}O_{(2)}$ bond. To calculate the nonequivalency of the oxygen atoms $O_{(1)}$ and $O_{(2)}$, first of all the disposition of water is fixed, and the position of the oxygen $O_{(1)}$ is changed along Z axis, and r_z and α values are varied for the value of the energy minimum obtained. In analog the structure of aquacomplex of alumoxane (boronoxane) without the proton is calculated. The energy $\Delta E_{theor}^{H^+}$ for the aquacomplex of alumoxane (boronoxane) is determined according to the following formulae:

$$\Delta E_{theor}^{H^+}(kJ\,/\,mole) = E_o(Al_2Cl_4O \cdot H_2O) - E_o(Al_2Cl_4O \cdot OH) \tag{10}$$

$$\Delta E_{theor}^{H^+}(kJ\,/\,mole) = E_o(B_2F_4O \cdot H_2O) - E_o(B_2F_4O \cdot OH) \tag{11}$$

and δpKa is determined from the correlation (9) or from the correlational ratio $\delta pKa = f(q_{H+})$ (4).

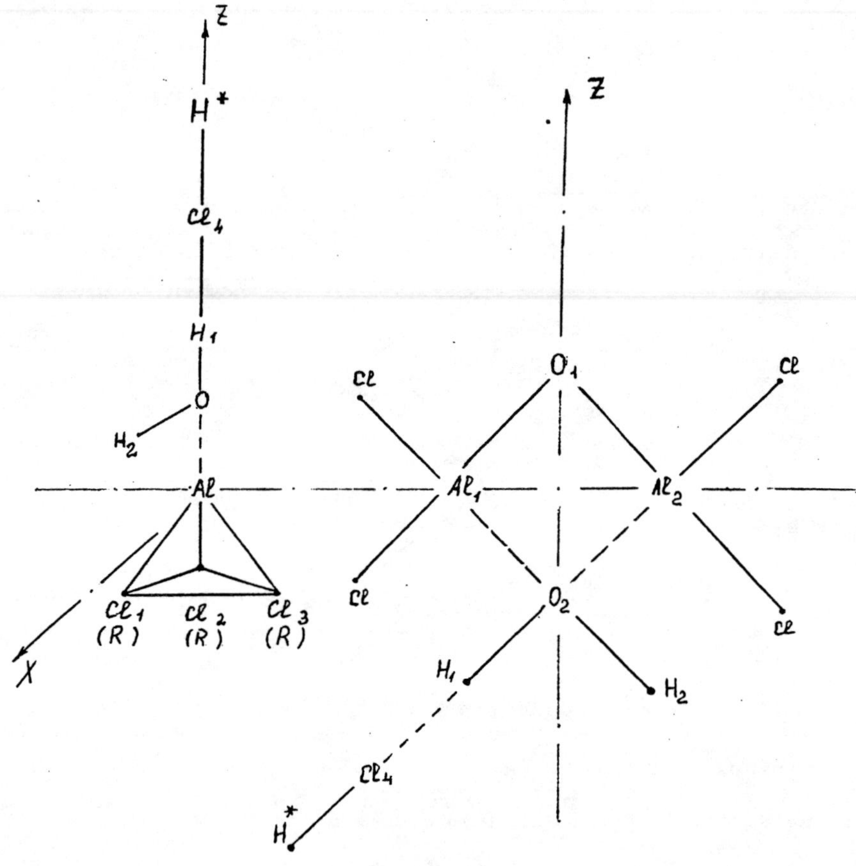

Figure 3.9. Models of triple complexes – composition $R_nAlCl_{3-n} \cdot H_2O \cdot HCl$.

The models of triple complexes $R_nAlCl_{3-n} \cdot H_2O \cdot HCl$ and $Al_2Cl_4O \cdot H_2O \cdot HCl$, the calculation of which is necessary in connection with taking into account the probable influence of HCl, being extracted as a result of possible transformations of $R_nAlCl_{3-n}\cdot H_2O$ under the influence of olefin [17, 19, 39], are determined on the Figure 3.9.

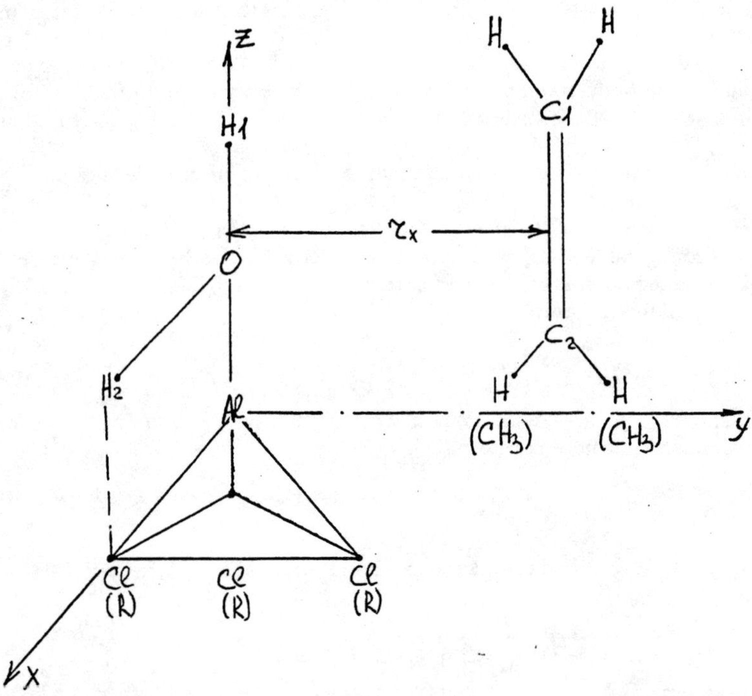

Figure 3.10. The initial models of triple complexes – composition Al-H₂O-olefin.

Calculation of the triple complexes $R_nAlCl_{3-n}(R_nBF_{3-n}\cdot R_nMgCl_{3-n})\cdot H_2O(R'OH, HCl)\cdot$olefin was performed in [19] by means of changing disposition of the olefin in relation to aquacomplexes with the determination of the equilibrium state r_x (Figure 3.10). Values of the bond lengths and angles between them for complexes $R_nAlCl_{3-n}\cdot(R_nBF_{3-n}\cdot R_nMgCl_{3-n})\cdot H_2O$ (HCl, R'OH) were corresponded to the energy without monomer. Water molecule was rigidly fixed by the account of coordination bond Al-O (B-O, Mg-O) and by bond O-H₂...Cl(O-H...F), being formed during the complex formation process [19].

The energy of interaction of aquacomplexes with the monomer and the energy of proton detachment $\Delta E^{H^+}_{theor}$ were determined according to the following formulae:

$$\Delta E^{H^+}_{B_3}\,(kJ\,/\,mole) = E_o(R_nAlCl_{3-n}\cdot H_2O\cdot M) - E_o(R_nAlCl_{3-n})\cdot H_2) + E_o(M) \qquad (12)$$

$$\Delta E^{H^+}_{theor}(kJ\,/\,mole) = E_o(R_nAlCl_{3-n}\cdot H_2O\cdot M) - E_o(R_nAlCl_{3-n}\cdot OH\cdot M) \qquad (13)$$

Studying the dynamics of interaction of the systems $R_nAlCl_{3-n}\cdot(R_nBF_{3-n}\cdot R_nMgCl_{3-n})\cdot H_2O$ (HCl, R'OH)·monomer the initial distance r_x from aquacomplexes to the monomer was taken equal 0.300-0.400 nm. Further on it was selected the reaction coordinate and parameters of optimization (bond lengths, angles). Monomer molecule was approached to the aquacomplex with the definite step by the reaction coordinate (Hook-Jives optimization method and (or) the method of variable metric). The process of the interaction of the monomer with the aquacomplex was conditionally subdivided into three stages: olefin coordination, π-bond break of olefin and the formation of the end structure.

In analog it was studied the dynamics of the interaction of molecules in the system monomer and monomer-AC, formed on the stage of the interaction of $R_nAlCl_{3-n}\cdot H_2O$ complex with olefins.

It was comfortable to calculate Arrhenius's activation energy (energetic barrier) by the CNDO/2 method plus BO-BE (bond order – bond energy) [75].

For the following formula:

$$RA+HCR' \rightarrow RA\text{-}H\text{-}CR' \rightarrow RAH+CR'$$

Johnson and Parr suggested the generalized formula of the potential energy of the system along the reaction trajectory [75]:

$$U=D_{15}(1\text{-}n^P)\text{-}D_{25}m^q+D_{35}(nm)^{\gamma}B$$

where D_{15}, D_{25}, D_{35} – bond energies; n, m - bonds order along the reaction trajectory, for example

$$(AlCl_3 \cdot H_2O)O \xrightarrow{\ m\ } H_1 \xrightarrow{\ n\ } C_1(C_4H_8)$$

$$
\begin{array}{ccc}
H & & H \\
 & \diagdown\ \diagup & \\
 & C_1 & \\
 & \| & \\
 & C_2 & \\
 & \diagup\ \diagdown & \\
CH_3 & & CH_3 \\
\end{array}
$$

R, q, γ should be taken from the corresponding tables [75].

$R_{CH} \approx 1.087$; $R_{OH} \approx 1.028$; $R_{OC} \approx \gamma$.

$\gamma = 0.26 \cdot \beta 3 \approx 0.26 \cdot 2.05 \approx 0.53$.

$\gamma\text{-}1 = -0.467$, $B = 1/2e^{-\beta 3 \Delta R_s} = |\ \Delta R_s = R_{CH}+R_{OH}-R_{OC} = 1.09+$

$+0.96\text{-}1.43 = 0.62\ | = 1/2e^{-2.05 \cdot 0.62} = 1/2e^{-1.27} = 0.14$.

R_{CH}, R_{OH}, R_{OC} - bonds lengths in nm, m=1-n;

$D_{1s} = D_{CH} = 105.5$ kcal/mole

$D_{2s} = D_{OH} = 114.7$ kcal/mole

$D_{3s} = D_{OC} = 85.4$ kcal/mole

$$
\begin{aligned}
U &= 105.5(1\text{-}n)^{1.087}-114.7(1\text{-}n)^{1.028}+ \\
&\quad +85.4 \cdot 0.14[n(1\text{-}n)]^{0.533} = 105.5(1\text{-}n)^{1.087}- \\
&\quad -114.7(1\text{-}n)^{1.028}+11.96[n(1\text{-}n)]^{0.533}
\end{aligned}
\tag{14}
$$

where n – the order of $H_{(1)}$-$C_{(1)}$ bond according to Armstrong [72] from the calculation of CNDO/2 method.

Testing U for extremum along the reaction coordinate Arrhenius's activation energy may be estimated.

To check the calculation and to compare the calculated data of complexes Fridail-Kraphtz acid (AlCl$_3$, BF$_3$, SnCl$_4$, etc.) – proton-donor (H2O, ROH, HCl, HF) it is comfortable to apply complexes of halogenides of alkaline metals with alkylaluminum-chlorides ($M^{+\delta}$, $RAlCl_3^{-\delta}$). The initial structures ($M^{+\delta}$, $RAlCl_3^{-\delta}$) were corresponded to the known structures $M^{+\delta}$, AlCl$_4$ [76]. It is sufficiently probable to substitute the ligand R at the aluminum atom by H atom in order to simplify the calculations. The anion $[HAlCl3]^{-\delta}$ was selected as tetrahedron with angles of 109° a t Al-Cl bonds lengths of 0.205 nm. Surely, the substitution of one of Cl atoms by H should principally lead to the change of angles and bond lengths of proper tetrahedron AlCl$_4$ [76]. But calculations have shown that in this case there occurs no sufficient change of electron density on other atoms of the complex.

Concerning the calculations of the most energetically profitable reaction trajectory of cations ($K^+=H^+$, Li$^+$, Na$^+$) of complexes $K^+HAlCl_3^-$ with olefin, in this case it is possible two directions of the cation attack (Figure 3.11).

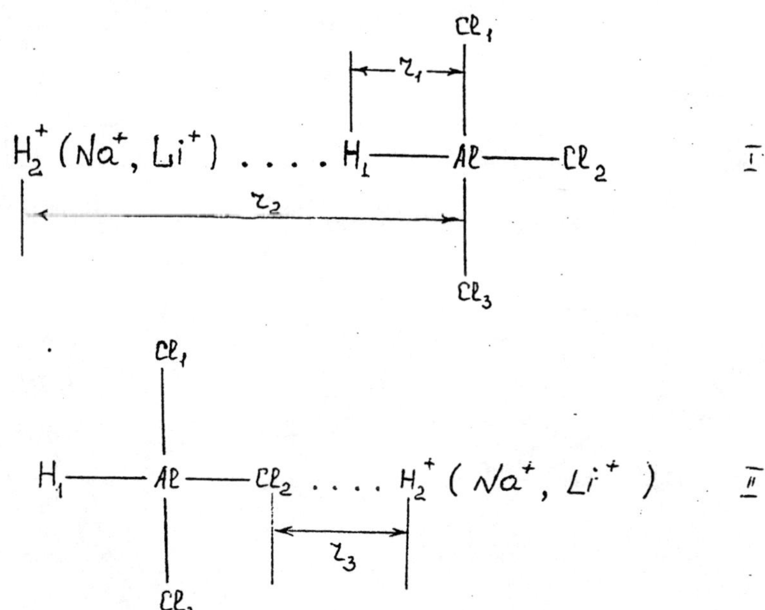

Figure 3.11. The models of complexes $M^{+\delta}$, $[HAlCl3]^{-\delta}$ (CNDO/2).

Stable state of the system (the minimum of the energy of the basic state) is defined by variation of distances r_1, r_2, r_3. In the case of the attack direction - 1, r_1 is changed first of all (from 0.164 nm up to 0.344 nm). For optimum value of r_1, changing r_2, there were obtained 10-15 values of charges on the atoms, dipole moments, system energies and othe quantum-chemical parameters. Then after graphical dependences of E_0 energy on r_2 value were depicted (at fixed value of r_1), and according to them equipotential energetic surfaces of the interaction of the corresponding molecules were built up. In the case of the attack of a proton by anion according to the direction 2, and also at the interaction of Na^+ and Li^+ with $[HAlCl_3]^-$ by both directions, r_1 was fixed at its optimum value, and minimum energy was determined, changing r_3 coordinate. Energies of the intersection ΔE_1 of cations with anion $[HAlCl_3]^-$ were calculated according to the formula:

$$\Delta E_1 (kJ/mole) = E(K^+HAlCl_3^-) - E(HAlCl_3^-) \tag{15}$$

To estimate the energy ΔE_2 of interaction of cations K^+ (H^+, Na^+, Li^+) with isobutylene the calculation of the model was performed, presented on Figure 3.12. In this case the distances r_4 and r_5 were optimized, and ΔE_2 value was calculated according to the following formula:

$$\Delta E_2 (kJ/mole) = E(K + C_4H_8) - E(C_4H_8) \tag{16}$$

The results of the calculations performed were compared with the literature data (calculated and experimental).

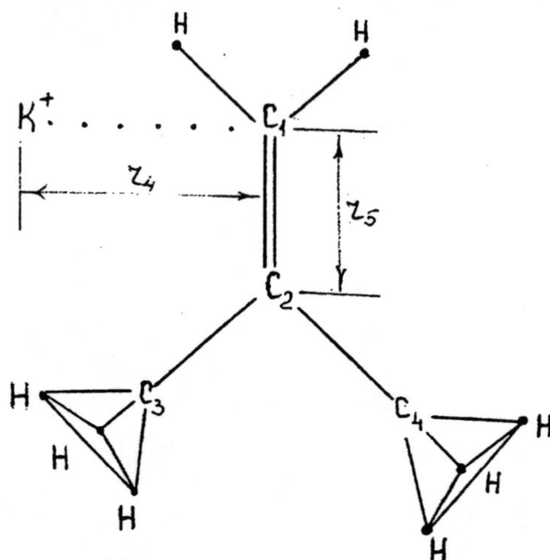

Figure 3.12. The model of interaction K^+ with isobutylene (CNDO/2).

The study of the interaction of $R_nAlCl_{3-n}\cdot H_2O(R'OH)$ complexes with the monomer was performed by the investigation of the structure of the potential energy surface of the system. Geometry of the initial state was set in accordance with the known (literature) data. The task of searching the optimum transitional structures is reduced to the search of the extremum points on the surface of potential energy. However, its solution for multiatomic systems is connected with practically unreal spenditure of the computer time. That is why it is inadvisable to apply the methodics of simplified calculations for such systems [77], where the calculation of all internal coordinates equal 3N-6 (N - the number of the atomic nuclei) is not strictly required, because the change of geometric parameters of the complexes fragments, disposed at the long distance from the reactionary centre and being not participate directly in the reaction, will influence insufficiently (in the ranges of the mistakes of the method) the character of the potential energy surface. Thus, the optimization of 70-80 parameters is reduced to the optimization of 5-10 parameters. The calculations of potential surfaces for some other chemical reactions are published in [77-79].

The calculation of the models, based on Fridail-Kraphtz acids with proton-donors in presence and in absence of a monomer, was performed as a rule without taking into account the influence of the solvent, i.e. in the approximation of isolated molecule. Such approach does not decrease the meaning of modelling of complex catalysts, because they are active in indifferent hydrocarbonic media, where solvatational effects are found insufficient.

3.1.2. MINDO/3 method.

According to MINDO method double-electron Coulomb integrals are calculated by the following modified Onho formula [64]:

$$\gamma_{AB} = \frac{1}{\sqrt{R_{AB}^2 + 0.25\left(\dfrac{1}{\gamma_{AA}} + \dfrac{1}{\gamma_{BB}}\right)}}$$

Resonance integral $\beta_{\mu\nu}$ is calculated by the formula according to the approximation of Malliken with the addition of empyric nondimensional multiplier G_{AB}, characterizing the type of atoms considering (AB bond):

$$\beta_{\mu\nu} = S_{\mu\nu}(J_\mu - J_\nu)G_{AB}$$

where J_μ, J_ν – ionization potentials;

$S_{\mu\nu}$ – the matrix of overlapping integrals;

A, B – types of the atoms;

μ, ν – orbitals indexes;

μ – A, ν – B;

EAB – the parameter, depending on the type of atom A or B.

Repulsion energy of the carcasses is estimated according to the following formula:

$$E_{carc} = \sum_A \sum_{<B} Z_A Z_B \left[\gamma_{AB} + \left(\frac{1}{R_{AB}} - \gamma_{AB}\right) \cdot \exp(-\alpha_{AB} R_{AB})\right]$$

where Z_A, A_B – carcass charge of A and B atoms;

γ_{AB} – integrals of electron repulsion;

R_{AB} – internuclear distance;

α_{AB} – an empyric parameter, characterizing types of the atoms under consideration.

The form of the ratio [18] is defined by the fact, that at low interatomic distances the repulsion energy of the carcasses is approximated well by the interaction of the point charges:

$$E_{carc} = \sum_A \sum_{<B} \frac{Z_A Z_B}{R_{AB}},$$

and at higher R_{AB}, it is the expression

$$E_{carc} = \sum_A \sum_{<B} Z_A Z_B \gamma_{AB}$$

Average mistake in the heats of the stable molecules atomization according to MINDO/3 method is 506 kcal/mole, which is lower than in calculations *ab initio* with the minimal basis. If the search for geometrical structure of the lowest energy (geometry optimization) is included into the calculation, the advantages of the semiempyric methods increase much higher and sufficiently sharp according to the above mentioned facts [64, p. 356].

Table 3.3 shows the values of relative energies of the protonated cyclopropanes, calculated according to MINDO/3 and *ab initio* methods with different bases.

MINDO/3 method reproduces the formation heats better than *ab initio* method even with the sufficiently wide basis of 6-31 G and allows to obtain the activation energies of the particular reaction with the mistake, similar to the determination of the formation heats. Morever, in spite of the fact that this method is parametrizated for the main states of the molecule, it is successfully applied also for the calculation of energies of the lowest singlet and triplet excited states [64].

To perform calculation of a molecule it is set experimental (initial) geometry of the object, the superposition of the symmetry conditions and functional dependences between the elements of the object geometry being probable. Experimental geometry of a molecule must be built in Cartesian coordinate system XYZ according to the following rules [80, 81]:

i) the disposition of each atom is defined by three coordinates: bond length, valent angle and torsional angle;

ii) first three atoms always lie in XY plane surface;

iii) first atom is always dislocated in the beginning of the coordinate system and is defined by its atomic number only;

iv) second atom is always dislocated on the positive direction of X axis and is defined by its atomic number and the bond length with the first atom;

v) third atom must dislocated in Xy plane surface in its first quadrant, i.e. between positive direction of X and Y axises. It is defined by the atomic number, bond length with the second atom and valent angle, formed by bonds between the third and the second ones, and between the second and the first atoms also;

vi) forth atom is defined according to the first three ones (using all three coordinates: bond length, valent and torsional angles);

vii) next atoms are set according to any three, defined before atoms;

viii) if it is difficult to determine the first three atoms as the ones of the calculating molecule, it is necessary to introduce the fictive atoms (atomic number 99); fictive atoms may be included always when this simplifies building up the experimental geometry.

It may be introduced any amount of the fictive atoms in condition if the total number of the atoms in geometry is not higher than 50. The introduction of experimental geometry is performed according to the atomic numbers sequence.

Table 3.3. Relative energies of the protonated cyclopropanes, kJ/mole [64].

No.	Structure	MINDO/3	Experiment	STO-3G	STO-4-31G	STO-6-31G
1	$CH_3CH_2CH_2^+$	77.79	75.28	82.39	70.68	58.90
2	CH_2 $H_3C \quad CH_2$ H^+	31.35	29.26	113.32	113.32	71.88
3	$H_2C = CH_2 \rightarrow CH_3^+$	51.04	29.26	95.39	72.31	54.34
4	$CH_3C^+HCH_3$	0.0	0.0	0.0	0.0	0.0

If the task is set about the optimization of the object geometry, and it is also known the condition of object geometry, it is recommended to set the symmetry conditions, which are expressed as the equilibrium of bond lengths, valent or torsional angles. The calculation of symmetry provides its adherence during all the process of optimization, including final object geometry. It is sufficient, that symmetry conditions decrease the amount of optimization parameters and, consequently, accelerate the process. Symmetry conditions may be superposed on optimizing geometry of the object, before calculation of functional dependences as well as after it. The optimization of geometry of the object is performed according to Davidone-Fletcher-Powel scheme [80].

Thus, semiempyric quantum-chemical methods CNDO/2 and MINDO/3 are the most suitable for the study of cationic polymerization of olefins in the presence of complex Lewis and Brenstedt acids taking into account the modern state of the computer technique (even comparing with nonempyric *ab initio* methods). In this connection it is logic to begin the analysis of practical quantum-chemical calculations.

CHAPTER 4

QUANTUM-CHEMICAL CALCULATIONS OF COMPLEX LEWIS AND BRENSTEDT ACIDS AS ACTIVE CENTRES OF CATIONIC POLYMERIZATION OF OLEFINS

Systematic quantum-chemical calculations of complex Lewis and Brenstedt acids in dependence on the nature of ligand surrounding were not practically performed. Just some calculations of the compositions of B, Al and some other elements with halogen- and oxygen-containing electrodonor agents were performed by both nonempyric and semiempyric methods with different methodics of optimization (by all parameters, if the object contains small amount of orbitals or fragmentally, if the amount of orbitals is high).

In particular, [82] performs nonempyric calculation by the method of CCP MO LCAO geometry and electron composition of HBF_4 model in comparison with HF and BF_3. Symmetry plane FBF was fixed at the calculation and the rest of geometric parameters was optimized. The acidity of compounds was studied on the basis of the obtained charge distribution and the change of electrostatic potential in the direction of F-H bond in molecules of HF (V_{HF}) and HBF_4(V_{HBF_4}). V_{HF} is lower than V_{HBF_4} in the range of the change of F-H bond lengths from 0.2 to 0.6 nm, which corresponds to higher electrical conductivity of HBF_4. This conclusion is also proved by direct calculation of the energy of a proton detachment from HF molecule. Complex HBF_4 acid is also considered in [83], where the surfaces of potential energy were calculated for different variants of proton approaching BF_4 anion in tetrahedral configuration taking into account both electron energy of the system and the Coulomb energy.

Dissociation of HBF_4 complex proceeds easier than that of HF. This correlates with the data on the initiation of cationic polymerization of olefins by the corresponding acids. In the ranges of Hartri-Fock-Rutane nonempyric method there were performed the calculations of geometrical composition and energies of different trajectories of decomposition of complex molecules and ions of $BcBH_4$, HBeB, $Mg(BH_4)_2$ [82]. There were considered two cases of intramolecular reactions. In the first case cation rotates among the anion preserving the structure of the latter, in the second case hydrite ion migrates with the decomposition of the anion structure. The conclusion was made about (dynamic) kinematic stability of anion in relation with intramolecular regrouping of the compound.

Moreover, it was performed quantum-chemical study of donor-acceptor $(CH_3)_2O \cdot BH$ complex by nonempyric MO LCAO CCP method [85]. The value of $R_{OB}=1.654 Å$ and the angle $\varphi(BOC)=152.8°$ were calculated for the optimized structure of this complex, the rest of geometrical parameters are the same as in literature. These values are calculated in agreement with the data, obtained by the method of electron diffraction. The analysis of separate contributions into the energy of complex formation showed, that the maximum component is corresponded to electrical interaction, which is twice stronger than the contributions of other linking interactions, and equals to the exchange energy with opposite sign. Surfaces of potential energies of benzene complexes with HF and HBF_4 were calculated by the CCP MO LCAO method [24].

Activating action of BF_3 is connected with the abrupt decrease of the reaction barrier for benzene with HBF_4. It follows from the calculation of energetic parameters of an intermediate state and separate compound (F, HF, BF_3, BF_4^-, HBF_4, C_6H_6), that the difference in energies between the main and transition state is 770 kJ/mole for the reaction $E(C_6H_7F)$ - $E(C_6H_6)$ - $E(HF)$ and 315 kJ/mole – for $E(C_6H_7BF_4)$ – $E(C_6H_6)$ – $E(HBF_4)$. The calculation showed that in both cases the reaction proceeds smoothly through the potential barrier forming no stable intermediate products.

The influence of the electro-donating ability of substitutors on the rate of processes is studied, comparing reactivity of alcoxyfluorborates with the data of quantum-chemical calculations for the determination of electron density in anions BF_3OR^- ($R=C_2H_5$, C_3H_7) in the ranges of MPDP Dewar method and complete optimization of structural parameters [86]. It was studied the kinetics of alkaline hydrolysis of BF_3OR^- anions. The reaction mechanism is defined by solvalytic dissociation of B-F bond. The rate of the reaction is limited on the stage of substitution of the first ion of boron fluoride. In the sequence

$$BF_4^- \rightarrow BF_3OCH_3^- \rightarrow BF_2(OCH_3)^- \rightarrow BF(OCH_3)_3^-$$

the charge q_B on the boron atom consists 0.37, 0.34, 0.30, 0.28, respectively. This allows to compare q_B with the strength of B-F bond, because the charge on fluorine atom is practically similar for all the calculated compounds (-0.32÷-0.35) due to high electrical negativity. That is why the value of q_B reflects, in fact, the energy of B-F bond. The increase of the reaction rate in the sequence of polysubstituted compositions of boron is connected, mainly, with the relative growth of the reaction barrier base. Whereas in the case of the above considered sequence of monosubstituted compositions of boron it is defined completely by the decrease of the energy of the transition state (Table 4.1).

Table 4.1. Basicity of substitutors pK, value of the activation energy $\Delta S^{\neq}$ and the rate of fluoride substitution in BF_3L^-.

No.	Anion	$\Delta S^{\neq}$, J/mole·K	pK, substitutor basicity	E, eV	K, min (at 273K)
1.	BF_4^-	144±2	3.2	18.36	$6.44 \cdot 10^{-7}$
2.	$BF_3 \cdot OCH_3^-$	94±2	15.1	14.21	0.029±0.001
3.	$BF_3 \cdot OC_2H_5^-$	119±2	15.9	13.50	0.051±0.002
4.	$BF_3 \cdot OC_3H_7^-$	–	16.1	12.82	0.073±0.002

As an example, illustrating the main regularities of donor-acceptor interactions at the application of aluminum compounds, we can take the calculation of ammonium bromide complex with dialkyl ester [87]. It was applied the method of estimation of the bond ionic character degree, the sense of which is displayed by molecular differential diagrams. In the end state I the charge on Al atom equals +0.111 and on the oxygen atom - +0.22.

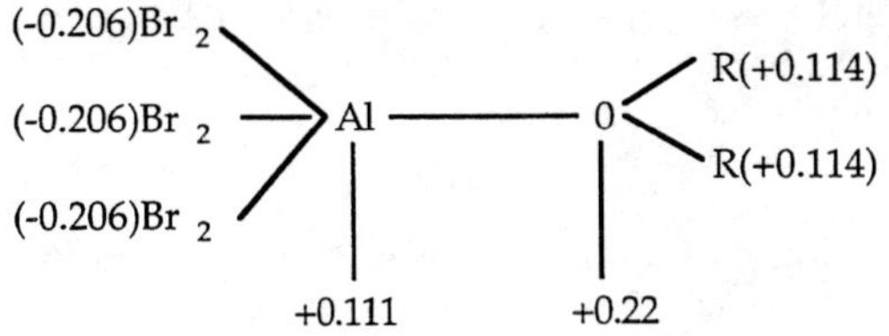

$$(-0.206)Br_2 \diagdown \quad \diagup R(+0.114)$$
$$(-0.206)Br_2 \longrightarrow Al \longrightarrow O$$
$$(-0.206)Br_2 \diagup \quad \diagdown R(+0.114)$$
$$+0.111 \qquad +0.22 \tag{I}$$

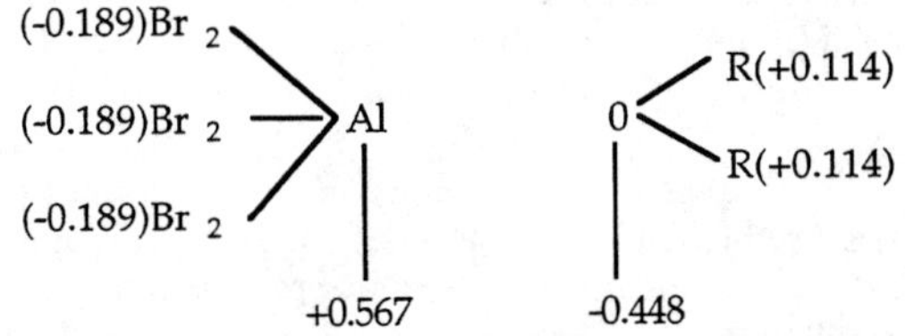

$$(-0.189)Br_2 \diagdown \quad \diagup R(+0.114)$$
$$(-0.189)Br_2 \longrightarrow Al \qquad O$$
$$(-0.189)Br_2 \diagup \quad \diagdown R(+0.114)$$
$$+0.567 \qquad -0.448 \tag{II}$$

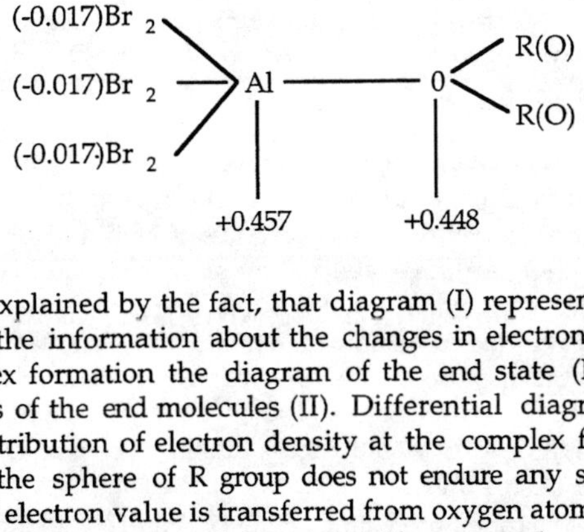

$$(-0.017)Br_2 \diagdown \quad \diagup R(O)$$
$$(-0.017)Br_2 \longrightarrow Al \longrightarrow O$$
$$(-0.017)Br_2 \diagup \quad \diagdown R(O)$$
$$+0.457 \qquad +0.448 \tag{III}$$

This result is explained by the fact, that diagram (I) represents the end state of the system. To obtain the information about the changes in electron structure of donor and acceptor at complex formation the diagram of the end state (I) should be compared with the diagrams of the end molecules (II). Differential diagram (III) characterizes changes in the distribution of electron density at the complex formation. In this case donor molecule in the sphere of R group does not endure any sufficient changes. The charge of 0.4-0.5 of electron value is transferred from oxygen atom to Al atom. The main part of this charge is left in the sphere of the Al atom, and only a small part of it is transferred to Br atoms.

Paper [88] performed the analysis of ionic pairs – models of cationic AC of $R^+MeX_{n+1}^-$ type (R=H, CH$_3$, OH; Me=B, Al; X=F, Cl) applying two methods of calculation – CNDO/2 and EHT. Both methods, adding each other, allowed to determine the dependence of the structure energies and charges on cations on nature of complexes and their configurations. Counterions retain tetrahedral structure at the interaction with cation, i.e. the models of AC selected reflect in a particular sense the nature of intermediate particles in the processes of cationic polymerization. It was shown low probability of esterates ionization, which testifies low value of the acidic strength of BF$_3$·R$_2$O complexes [89]. The increase of negative charge in ligands, linked with the central atom of the acceptor, leads to the increase of polarity of Al - ligand bonds, which is concorded with the increased activity of ligands in nucleophilic processes [18].

It should be mentioned, that the value of the charge transferred and the character of its distribution in one and the same complex depends weakly on the applied method of calculation [96]. It is shown simultaneously, that quantum-chemical calculations of complexes of Lewis and Brenstedt acids allow to obtain the information about their structure and energetic characteristics as well as to estimate their acidity and the reactivity of compounds, which is connected with it.

Correlation of calculated values of the proton detachment energies and the charge on hydrogen atom with the experimental ones of the universal index of acidity is set for unitypical sequences of acids [91, 4]. The correlation shown may be applied for the analysis of the experimental data. For example, paper [73, p. 100] shows the investigation of trimethylaluminum aquacomplex by CNDO/2 method. The conclusion of low Brenstedt acidity of this complex was made, basing on the value of the complex formation energy (309 kJ/mole) and the charge on hydrogen atoms of water (0.2 atomic unit). As the acidity strength of catalysts defines their catalytic properties in the processes of polymerization, the activity of $AlR_3 \cdot H_2O$ complex at styrene polymerization [2, p. 34] testifies low probability of the initiation by Brenstedt acid.

Similar methodics of the analysis was used for interpretation of the results of calculation by CCP MO LCAO method in valent approximation CNDO/2 of $AlCl_3 \cdot H_2O$ donor-acceptor complex [19]. In absence of additional polarizing factors the complex is stable and the proton detachment from water molecule is unprofitable energetically. The comparison of the energy of proton detachment with the experimental data of electrophilic activity of the complexes, based on $AlCl_3 \cdot H_2O$ and $RAlCl_2 \cdot H_2O$, point out the necessity to taken into account additional factors at proton attachment from aquacomplexes.

The values of charges on hydrogen atoms of water, according to which it is possible to estimate the acidic strength of a complex [4] on the surface of SiO_2 are shown in the papers [92, 93]. Electron structure and energetics of the formation of adsorptional complexes of water on SiO_2 surface, stabilized by hydrogenic and coordinational bonds, was studied by the MINDO method. Stabilization of adsorptional structures by means of several hydrogen bonds does not allow to explain the observed high values of the initial heats of water adsorption on the dehydrated surface of silicon dioxide. The formation of coordinational complexes of H_2O is more profitable energetically. This gives the basis to consider hydroxylated atoms of silicon on the surface of SiO_2 as the initial centers of water absorption.

Heterolythic dissociation of water $H_2O \rightarrow OH^- + H^+$ requires higher energy expenses [94]. That is why in usual conditions H_2O molecules possess no acidic properties. however, the acidity of H_2O is displayed as a result of loosening up O-H-bond in the complexes with R_nAlCl_{3-n}, SiO_2, etc. or in the field of cations of metals, especially multicharged ones. This is pointed out by the values of negative logarithms of the first constants pKr of hydrolysis, which are lower than pKa of water (15.7) and decrease at the transition from single-charged to three-charged cations:

$$[Na(H_2O)_x]^+ \rightarrow [NaOH(H_2O)_{x-1}]^+ \ (pKr = 14.6),$$

$$[Mg(H_2O)_x]^{2+} \rightarrow [Mg(OH)(H_2O)_{x-1}]^+ \ (pKr = 11.4),$$

$$[Al(H_2O)_x]^{3+} \rightarrow [AlOH(H_2O)_{x-1}]^{2+} \ (pKr = 5.0) \ [95].$$

Interesting correlational equation

$$E_r = a + b\varphi$$

$(E_r$ – energy of O"-H bond; a, b – constant coefficients; φ – valent angle of carbonic acids $118° < \varphi < 130°)$ was obtained with the help of CNDO/2 method. The correlation allows to calculate constants of ionization of carbonic acids in one solution, based on the application of the known values of valent angle OCO in carboxyl group [96]. The equation may be applied for the determination of carboxyl groups of polycarbonic acids. There exists correlational dependence between pKr (pKr=lgK) and X-ray structural data on the composition of carboxyl groups in the molecules of carbonic acids.

In the number of works the mechanisms of some electrophilic catalytic reactions are suggested, basing on quantum-chemical calculations [92, 93]. In the paper [92] it was performed the calculation of electron structure and energy of proton detachment by CNDO/BW method for clusters of zeolite $(A'O)_{m-n}(A"O)_n Si(OH)-Al(OH)_3 m=3$, n=0, 1, 2, 3; A', A" – univalent atoms, modelling the interaction with Si and Al atoms in the structure of zeolite qualitatively right impart the increase of acidic properties of H-forms of zeolites at the increase of Si and Al ratio. In the ranges of the selected model the mechanism of the double bond transfer is supposed in the reaction of isomerization of 1-butene into 2-butene on acidic centres of Brenstedt type. Quantum-chemical calculations are adjusted with spectroscopic data, which allow to compare the acidity of different hydroxyl groups not with the help of specially selected indicators, but by means of the analysis of the charges distribution on the reacting atoms themselves, even if this concerns the excited or intermediate states. Thus, it is disclosed the possibility of direct determination of thermodynamic and kinetic parameters of the catalyst acidity [93]. The data presented allowed to formulate spectroscopic criterion of the coordinate of chemical reactions, proceeding on Brentedt's acidic centers. In the case, if the activation energy calculated theoretically (basing on quantum-chemical calculations or on the frequencies of valent oscillations of O-H bond) is close to its experimental value, the main contribution is made by the proton transfer from catalyst to the reacting molecule. The reaction proceeds by the stage acidic mechanism. If these values differ sufficiently, then the mechanism must be coordinated or completely disconnected with the acidic properties of the surface.

The results important from the point of view of the alcohols behavior in cationic reactions were obtained at the interaction of H^+ and hypothetic Brenstedt acid $H-O^+$ with C_2H_5OH [97, 98]. Basing on the charges on atoms, the part of transferred charge and relative changes of values of the chemical bonds energies the conclusion was made, that C_α-O bond of the system is weakened more sufficiently than any other bond at the interaction of alcohol with $H-O^+$ centers, i.e. the calculation testifies the benefit of carbcationic function of the alcohol.

Similar results were obtained at the study of the interaction of ethyl alcohol with alumosilicate catalysts by CNDO/BW method [99]. Basing on charges on atoms, bonds indexes:

$$i_{AB}' = \frac{E_{AB}^k - E_{AB}^c}{E_{AB}^c} \cdot 100\%,$$

(where E_{AB}^{k} and E_{AB}^{c} – double-centered components of energy of coordinated and free molecule), it was concluded that there proceeds the selective weakening of C_α-O and C_β-O bonds. The calculation testifies for the benefit of the probability of the coincident mechanism of cationic reaction of alcohol dehydration.

The dependence of the behavior of secondary protonated alcohols $H^+ROH(OH)CH_3$, where $R=CH_3$, C_2H_3, iso-C_3H_7, tret-C_4H_9 on nature of the radical was discovered in the paper [100]. The calculation of bonds degrees, charges on atoms and relative change of E_O of the system testifies the weakening of O-R with the increase of branching degree of R radical. The calculated data obtained correlate well with the experimental values of the rates of catalytic dehydration of the secondary alcohols on SiO_2 in dependence on their structure. Paper [101] shows the calculation of the interaction of ethyl alcohol with Lewis γ-Al_2O_3 acid by modified semiempyric CNDO/2 method. In difference to the above referred papers it was stated the proton-donor function of the alcohol. Analyzing the energy of the complex formation, indexes of the change of chemical bonds strengths, the change of charges on atoms, parts of proton transfer, the authors consider that O-H bond of the alcohol is responsible for physical adsorption of oxide on the surface. The interaction of C_2H_5OH with a list of metals (Ni, Cu, Al) was studied by Huckel method [102]. The values of the bonds degrees, correlating with their occupancy, show that for Ni the occupancy of bonds changes insufficiently, i.e. Ni displays no catalytic activity in relation to alcohol. In the case of copper O-H bond is weakened more sufficiently than others, and for Al this effect is expressed weaker. Consequently, ethyl alcohol is inclined to display proton-donor function.

Theoretical analysis of complexes of various Lewis and Brenstedt acids allows to conclude that the amount of the objects considered is comparatively small. And they are far from the models of AC of real catalytic systems, with some exceptions. Particularly, this relates to AC, which include a monomer in their composition, and also to the set of correlation between calculations and the experimental data.

Thus, the collection of the experimental and theoretical data existing does not allow to imagine the mechanism of excitation of cationic polymerization of olefins under the influence of Lewis and Brenstedt acids. The influence of a monomer on the behavior of complexes at excitation of cationic polymerization, the necessity of calculation of which follows from the experimental investigations, is not considered in theoretical works. It is specially concerned with weakly acidic chlorides and alkylchlorides of aluminum with the proton-donors such as water and alcohol. It should be pointed out also the absence of the attempts to generalize the results of theoretical investigations of the objects of different alcohol complexes, close by nature, and to compare them with the experimental data. It was succeeded to improve this failure, considering complex systems of halogenides of metals Al, Mg, B with proton-donor from quantum-chemical point of view.

4.1. Complexes of Hydrogen Chloride with Aluminum Chlorides.

Complexes of $R_nAlCl_{3-n}\cdot HCl$ type represent suitable object for quantum-chemical calculations, because the composition and structure of the complexes were proved experimentally (in the composition of triple $R_nAlCl_{3-n}\cdot HCl$+toluene complexes) [103], as well as their dependence on Al nature, and the activity (selectivity) of action of the complexes as catalysts in different cationic processes was also stated. Comparative calculations allow to refine the notion about the nature of action of complex catalysts as well as to put forward the question about the prediction of catalytic activity in the number of classic electrophilic reactions.

In the general case the mechanism of catalytic action of complexes of hydrogen chloride with aluminum chlorides is relatively simple and understandable [9]. That is why the improvement of the experimental behavior of $Cl_3Al\cdot HCl$ complexes by the calculated data may be applied for verification of correctness of CNDO/2 method according to the set classic parameterization and for calculation of other complexes, based on aluminum chloride.

Calculated parameters of acidic strength of R_nAlCl_{3-n} complexes are presented in the Table 4.2. It is seen that for heterolytic break of the bond in HCl molecule the energy is required $\Delta E_{det\,ach}^{H^+} = -440$ kJ/mole. At the complex formation of HCl with aluminum chlorides positive charge on H atom increases from +0.09 to $+0.18\div+0.20$. The values of Δq_{H^+} (the part of charge growth on the hydrogen atom of HCl by means of its complex formation with R_nAlCl_{3-n}) are approximately similar for all the complexes ($0.09\div0.11$).

Table 4.2. Parameters of acidic strength of the models of complexes $R_nAlCl_{3-n}\cdot HCl$ (CNDO/2).

No.	Complex	ΔE^{H^+}, kJ/mole	q_{H^+}	Δq_{H^+}
1.	HCl	-440	+0.09	–
2.	$HCl\cdot(CH_3)Al_3$	+280	+0.18	0.09
3.	$HCl\cdot(CH_3)_2AlCl$	+280	+0.18	0.09
4.	$HCl\cdot OHAlCl_2$	+340	+0.20	0.11
5.	$HCl\cdot CH_3AlCl_2$	+260	+0.18	0.09
6.	$HCl\cdot AlCl_3$	+300	+0.19	0.10
7.	$HCl\cdot EtAlCl_2$	+270	+0.18	0.10

Proton detachment from $HCl\cdot R_nAlCl_{3-n}$ complexes is accompanied by the energy extraction of about 260-340 kJ/mole (in difference with single HCl). This process is energetically profitable. This fact is also proved by the experimental data on high acidic strength of $HCl\cdot R_nAlCl_{3-n}$ complexes [3], that testifies about the suitability of CNDO/2 method for calculating the considered models. The differences in the acidic strength of complexes are levelled, as judged from the values of $\Delta E_{det\,ach}^{H^+}$ and Δq_{H^+} in dependence on nature of Lewis acid in complexes. The independence of acidic strength of complexes on nature of Lewis acid displays that all complexes of HCl with aluminum chlorides are inclined to ionization according to the following reaction:

$$HCl + R_nAlCl_{3-n} \underset{\leftarrow}{\overset{\rightarrow}{=}} H^{+\delta}[R_nAlCl_{3-n}]^{-\delta}.$$

In its turn, this correlates with the literature data on the fact that all complexes of HCl with different Lewis acids from $AlCl_3$ display the activity in cationic polymerization of olefins [3]. Simultaneously, the obtained calculated data point out the advisability of direct initiation according to the reaction of ionized complex with a monomer instead of the scheme, suggesting preliminary interaction of HCl with R_nAlCl_{3-n} forming stronger Lewis acids. The data obtained correlate also with high activity as well as with low selectivity of action of the corresponding catalysts, based on HCl in the reaction of isobutylene polymerization (Table 4.3) [103].

Table 4.3. Polymerization of isobutylene from C_4 fraction hydrocarbons under the influence of $R_nAlCl_{3-n} \cdot HCl$ (Al-7.3$\cdot$10^{-3} mole/l, C_4 - 1.3 mole/l, CH_3Cl_3, 253 K, 6$\cdot$10^2 sec).

	Catalyst	Conversion of iso-C_4H_8%	Selectivity by iso-C_4H_8%
1.	$2AlCl_3 \cdot HCl \cdot 3C_6H_5CH$	99	0.36
2.	$C_5H_5AlCl_2 \cdot HCl \cdot C_6H_5CH_3$	86	0.43
3.	$(C_2H_5)_2AlCl \cdot HCl \cdot C_6H_5CH_3$	98	0.53

Note: the composition of C_4 fraction (mass %): C_3H_6 - 0.6; n, i-C_4H_8 - 55.1; i-C_4H_8 - 14.9; α, β-C_4H_8 - 29.5

Similar character of behavior of different HCl complexes follows from their structure.

$$CH_3 - \hexagon \; + HCl + R_nAlCl_{3-n} \underset{\leftarrow}{\overset{\rightarrow}{\ldots\ldots\ldots}} R_nAlCl_{3-n}$$

i.e. in all cases the complexes of the proton attachment are formed, which differ by nature of counterion. Simultaneously, the obtained calculation data for the complexes of $R_nAlCl_{3-n} \cdot HCl$ in combination with the experiment point out the necessity to perform the next step of the work – calculation, modelling the interaction of $AlCl_3 \cdot HCl$ complex with olefins (ethylene, propylene, isobutylene).

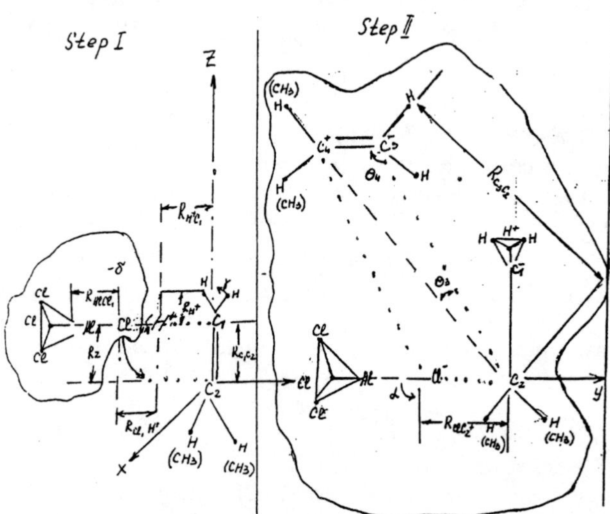

Figure 4.1. Mechanism of initiation at cationic polymerization of ethylene, propylene and isobutylene in presence of HCl-$AlCl_3$.

Quantum-chemical characteristics of the reaction of $AlCl_3 \cdot HCl$ with olefins are presented in the Table 4.4 The behavior of separate atoms and fragments of the system at such interaction is exhibited on the Figure 4.1. Optimization of structures according to the reaction coordinate r_{H+Cl} at the calculation by CNDO/2 and MINDO/3 methods was performed according to one and the same parameters: length of the olefin double bond ($r_{C_1C_2}$), proton H displacement along Z axis (r_{H+}), displacement of counterion $[Cl_3AlCl]^{-\delta}$ relative to olefin along Z axis (r_{H+}), the angle of forming tetrahedron C_1HH^+H of carbcation. The initial value of the reaction coordinate r_{H+Cl} was accepted equal 0.300 nm, and the energy of the model structure of the first stage – zero (all trajectory of the reaction is subdivided into 10 conditional stages of interaction (Table 4.4)).

As the particle $H^{+\delta}$ approaches olefin by the reaction coordinate r_{H+Cl} up to the 5-the stage, there occurs gradual loosening of π-bond of the olefin – the increase of the bond length $r_{C_1C_2}$ from 0.132 nm to 0.138 nm. In this case total energy of the system E_o decreases from 0 to -95 kJ/mole. The break of double bond begins on the 6-th stage ($r_{C_1C_2}$ increases from 0.138 nm to 0.146 nm). There appears sufficient decrease of the total energy of the system from 095 kJ/mole to -630 kJ/mole. $H^{+\delta}$ attacks the most hydrogenated C_1 atom of isobutylene ($r_{C_1C_2}$=0.148 nm), in this case the decrease of E_o value down to -870 kJ/mole is observed. Let us point out that for gas-phase reaction the affinity of proton to isobutylene ΔH_{298} is 820.6 kJ/mole [112].

Table 4.4 Optimization parameters, varied by the reaction coordinate $r_{H+C(1)}$, and some calculated ones (E_o, q_{H+}, q_{Cl}, q_{C1}, q_{C2}) at the interaction of $AlCl_3 \cdot HCl$ complex with isobutylene.

Stage	Reaction coordinate r_{H+C} (nm)	Optimization parameters nm			grad	F_{o}, kJ/mole	q_{H+}	q_{Cl}	q_{C1}	q_{C2}
		$r_{C_1C_2}$	r_{H+}	r	γ					
1.	0.300	0.132	0	0	0	0	+0.19	-0.20	-0.11	+0.05
2.	0.280	0.134	0	0	0	-31	+0.19	-0.21	-0.12	+0.07
3.	0.260	0.134	0.02	0	0	-48	+0.20	-0.23	-0.15	+0.10
4.	0.220	0.136	0.02	0	0	-73	+0.21	-0.24	-0.20	+0.15
5.	0.180	0.138	0.04	0	0	-95	+0.22	-0.24	-0.22	+0.19
6.	0.160	0.146	0.09	0.02	0	-630	+0.08	-0.19	-0.05	+0.22
7.	0.150	0.148	0.09	0.02	3	-710	+0.04	-0.20	-0.04	+0.21
8.	0.130	0.148	0.08	0.06	9	-805	+0.04	-0.21	+0.03	+0.22
9.	0.120	0.148	0.08	0.12	17	-845	+0.04	-0.22	-0.03	+0.20
10.	0.110	0.148	0.07	0.15	19	-870	+0.03	-0.21	-0.03	+0.20

It is seen from the Table 4.4 that total energy E_o of the system decreases infinitely along all consecutive stages of the reaction, i.e. the interaction of $AlCl_3 \cdot HCl$ complex with olefins, with isobutylene in particular, possesses nonbarrier character. In this case the behavior of $[Cl_3AlCl]^{-\delta}$ fragment is interesting (counterion been formed). Fragment

$[Cl_3AlCl]^{-\delta}$ does not participate in any interactions, except the ones, connected with the break of H-Cl bond. This situation exists practically up to the complete break of π-bond of olefin (up to 7-8 stage). At the detachment of H^+ proton from $AlCl_3 \cdot HCl$ complex additional interactions are also possible between $[Cl_3AlCl]^{-\delta}$ fragment and olefin (on the 8-10 stages). On the last stages of the reaction the formation of $[Cl_3AlCl]^{-\delta}$ counterion is finished, which together with carbcation $(CH_3)_3C^{+\delta}$ forms polarized intermediate compound $[Cl_3AlCl]^{-\delta}\ldots\ldots C^{+\delta}(CH_3)_3$, corresponded to AC (Figure 4.1).

Changes of charges on atoms H, Cl, C_1, C_2 of $AlCl_3 \cdot HCl \cdot C_4H_8$ system are presented in the Table 3.4. The charge q_{H^+} does not practically change at H^+ particle approaching isobutylene (more correct, it increases insufficiently from +0.19 to +0.22; CNDO/2 method mistake according to charges on atoms is ±0.02). According to correlational dependence pKa – q_{H^+} [4] this is corresponded to the increase of pKa from +7.7 to +4.4. In all probability, the invariability of the charge of H^+ particle in cationic initiation is characteristic for nonbarrier reactions under the influence of strong acids or "super"-acids. The estimated small contribution of Coulomb's (electrostatic) component into the energy of interaction of $AlCl_3 \cdot HCl$ complex with isobutylene testifies the fact, that such reactions are weakly connected with the absolute value of the reagents charges. The sufficient contribution can be made by covalent characteristic of H^+ (J - Q, J - ionization potential, Q - hydration heat) or by covalent term of its acidic-base interaction with olefin, possessing value degree similar to carbcation [112]. Let us point out, that electrostatic characteristic of H^+ cation (e^2/r, e - charge, r - radius) exceeds sufficiently its values for all known cations.

At the approach of $H^{+\delta}$ to β-carbonic atom C_1 of isobutylene the charge q_{H^+} decreases, and tetrahedron $C_1^-HH^+H$ of carbcation is formed. Positive charge on α-carbonic atom C_2^+ increases according to the reaction coordinate from +0.05 to +0.22 by means of displacement of π-electron density of double bond to C_1 atom. Atom C_1, although it receives the additional electron density, endures no sufficient changes resulting the reaction (Table 4.4).

The analysis of charges on atoms and fragments of $AlCl_3 \cdot HCl \cdot C_4H_8$ system shows, that proton transfer proceeds according to common scheme of its accepting by olefin. Consequently, the mechanism of initiation of cationic polymerization of isobutylene by strong acidic $AlCl_3 \cdot HCl$ complex is connected with competitive interaction of olefin and counterion in the reaction with proton. The delay attachment of the counterion formed to carbcation does not exclude the possibility of the formation of free carbcation, that takes place in some systems.

Studying the influence of counterion on the interaction of carbcation with olefin (the stage of the chain growth), the mostly stable configuration of the system ($r_{C_2C_3}$=0.30 nm, Figure 1) was determined by the olefin molecule rotation among C$_2$-Cl bond (rotation radii are $r_{C_2C_3}$=0.25 nm - 0.50 nm). Distance $R_{C_2C_1}$ was selected as the reaction coordinate. Optimization of the structure along the reaction trajectory was performed by the following parameters: angle of olefin rotation in the planes ZOX (Q_3, Q_4) and ZOY (φ_3, φ_4); double bond length ($r_{C_3C_4}$); angle of counterion rotation in the plane ZOX (α); length of the forming bond ($r_{C_2C_3}$). The behavior of separate atoms and fragments of the system at the interaction of AC-polarized intermediate compound

$[Cl_3AlCl]^{-\delta}$......$C^{+\delta}(CH_3)_3$, formed on the stage, with isobutylene and the change of charges on atoms of AC testifies the coordinated character of the reaction of carbcation with olefin, similar to the ones described in [104]. The reaction of carbcation with olefin represents unusual combination of electrostatic and orbital interactions, controlled by counterions. Making the approach to carbcationic center difficult on the coordination stage, counterion provides simultaneously the direction of the further attack, which is displayed in the mutual orientation of the system components. The reaction is characterized by energetic barrier 34 kJ/mole high and by the energy gain of 617 kJ/mole as a result of AC reaction with isobutylene (see Table 4.5).

Table 4.5. Quantum-chemical characteristics of $HCl \cdot AlCl_3$ interaction with olefin.

No.	System	Initiation		Chain growth	
		CNDO/2	MINDO/3	CNDO/2	
		E_b kJ/mole	E_b kJ/mole	E_a kJ/mole	E_b kJ/mole
1.	$AlCl_3 \cdot HCl \cdot C_4H_8$	870	356	34	617
2.	$AlCl_3 \cdot HCl \cdot C_3H_6$	858	327	61	583
3.	$AlCl_3 \cdot HCl \cdot C_2H_4$	823	294	82	566

The value of E_b in the sequence of olefins C_4H_8, C_3H_6, C_2H_4 decreases on both stages of initiation and chain growth (For the first stage it was displayed by two methods CNDO/2 and MINDO/3). Energetic barriers in this sequence increase, that is in agreement with cationic activity of the olefins studied (Table 4.5) [2].

4.2. Complexes of Hydrogen Fluoride with Boron Fluorides.

Electron structures and geometry of $HFBR_nF_{3-n}$ complexes are close to the structure of HBF_4 by their characteristics [55] and possess the form of distorted tetrahedra. The data of quantum-chemical calculations of the complexes are presented in the Table 4.6.

Table 4.6. Tetrahedra of acidic strength of the models of $HF \cdot R_nBF_{3-n}$ complexes*.

No.	Complexes	q_{H^+}	pKa
1.	$F_3B \cdot HF$	+0.50	-60
2.	$C_3BF_2 \cdot HF$	+0.49	-59
3.	$(CH_3)_2BF \cdot HF$	+0.48	-58
4.	$(CH_3)_3B \cdot HF$	+0.48	-58
5.	$C_2H_5BF \cdot HF$	+0.49	-59
6.	$(C_2H_5)_2BF \cdot HF$	+0.48	-58
7.	$(C_2H_5)_2B \cdot HF$	+0.48	-58

* - for R=iso-C_3H_7 and tret-C_4H_9 values of q_{H^+} and pKa are situated in the same range.

The values of q_{H^+} are positive and very high, and universal indexes of acidity pKa possess high negative values, respectively, i.e. $HF \cdot BR_nF_{3-n}$ are corresponded to strong acids.

In this case values of pKa depend practically neither on ligand surrounding of B nor on nature of the ligand radical. The proton detachment from $R_nBF_{3-n} \cdot HF$ is energetically profitable and is accompanied by extraction of the energy (higher than 350 kJ/mole). This correlates with the experimental data on high acidic strength of the complex HBF_4 acid [3]. In this connection it should be expected, as in the case of strong acidic $R_nAlCl_{3-n} \cdot HCl$, that the process of initiation of cationic polymerization of olefins of HBF_4 complexes will possess nonbarrier character and represent usual proton acceptance of olefin. In particular, this is shown on the example of modelling of interaction of $BF_3 \cdot HF$ complex with isobutylene.

4.3. Complexes of Alkaline Metals.

The difference in the nature of ligands of aluminum and boron does not influence sufficiently the acidic strength of compounds of the calculated complex acids $R_nAlCl_{3-n} \cdot HCl$ and $R_nBF_{3-n} \cdot HF$. In general case the nonsymmetry of the ligand surrounding influences the behavior of complex acids. The character of influence at the definite ligand surrounding will depend on nature of the external-spheric cation.

In opposite to rigid, symmetric anions – $AlCl_2(BF_4)$, which are relatively stable in the processes of polymerization of olefins, nonsymmetric counterions of $R_2AlCl_2^-(R_2BF_2^-)$ type contain alkyl radicals, more mobile than halogen, which are detached at electrophilic attack of the anion [105]. On the other side, the complexes $M^{+\delta}RAlCl_3^{-\delta}$ behave theirselves in another way: at the cationic attack counterion decomposes with Cl ligand detachment instead of R one [106].

In this connection the activity of cations was studied theoretically on the example of $M^{+\delta}HAlCl_3^{-\delta}$ in the competitive reactions with counterion fragment and with isobutylene. The results of calculations characterizing the interaction of cations with $H_{(1)}AlCl_3^-$ anion are presented in the Table 4.7. They testify the nonequality of the reaction, proceeding in directions I and II at the attack of ligands $H_{(1)}$ and Cl by cations (H^+, Na^+ and Li^+) (Figure 3.11). There is formed no stable structure at the attack by $H_{(2)}^+$ proton according to the II direction of the reaction. In the case of the proton approaching $H_{(1)}$-Al bond of the anion the system is characterized by the mostly deep potential well, possessing $E_0=0138440$ kJ/mole, at $r_1=0.264$ nm. In this case all signs of the reaction proceeding are observed. Figure 4.2 presents equipotential energetic surface. The closer atoms $H_{(2)}$ and $H_{(1)}$ are to Al atom and to each other, the sharper the surface front is. The far they are, the smoother the front is. Table 4.7 contains also other designations on $H_{(2)}^+$ interaction with $H_{(1)}AlCl_3^-$ fragment. As it was expected, at r_1 change from 0.43 nm to 0.29 nm dipole moment of the system decreases, positive charge on Al atom increases, and negative charges on $H_{(1)}$, $Cl_{(1)}$ and $Cl_{(2)}$ atoms and positive charge on H atom possess the tendency to decrease.

Cations Na^+ and Li^+ are less selective in relation to ligands H and Cl of anionic fragment, although the reaction by $Al-H_{(1)}$ bond is more profitable for them (Table 4.7). The decrease of r_2 and r_3 distances is at first accompanied by the growth of general energy of the system, and then by its decrease. The values of E_{bond} for $Na^+ - H_{(1)}AlCl_3^-$ and $Li^+ - H_{(1)}AlCl_3^-$ systems change similarly.

Dipole moments of the systems are also sensitive to the change of r_2 and r_3 coordinates. The increase of dipole moments values for coordinates, corresponded to the maximum of $E_{K^+H_{(1)}}AlCl_3^-$ – at the transition from H^+ to Na^+ and further to Li^+ $K^+H_{(1)}$ (2.35; 5.90 and 6.46 D), relates to the weakening of interactions of the pointed out cation with the anion ($\Delta E_{(1)}$ values equal 1920, 1250 and 890 kJ/mole respectively, see the Table 4.7). Characteristics of charges of the directly reacting atoms: Na, Li and $H_{(1)}$ – for attack in the direction I, and Na, Li and Cl – for attack in the direction II – testify the tendency of different signed charges to annihilation. It should be mentioned, that although the tendency to the increase of positive charge on Al (with the increase of r_{r2} coordinate) is preserved for $Na^+H_{(1)}AlCl_3^-$ and $Li^+H_{(1)}AlCl_3^-$, in some cases central atom of the anion is found negatively charged (Table 4.7). This effect is connected with the influence of d-orbitals and is formal in some sense. Thus, metal-hydride bond of $H_{(1)}AlCl_3^-$ anion is the most vulnerable place at the electrophilic attack. Consequently, it may be supposed, that Al-R bond will be the most vulnerable place for $M^{+\delta}RAlCl_{3-n}^{-\delta}$ complexes at electrophilic attack.

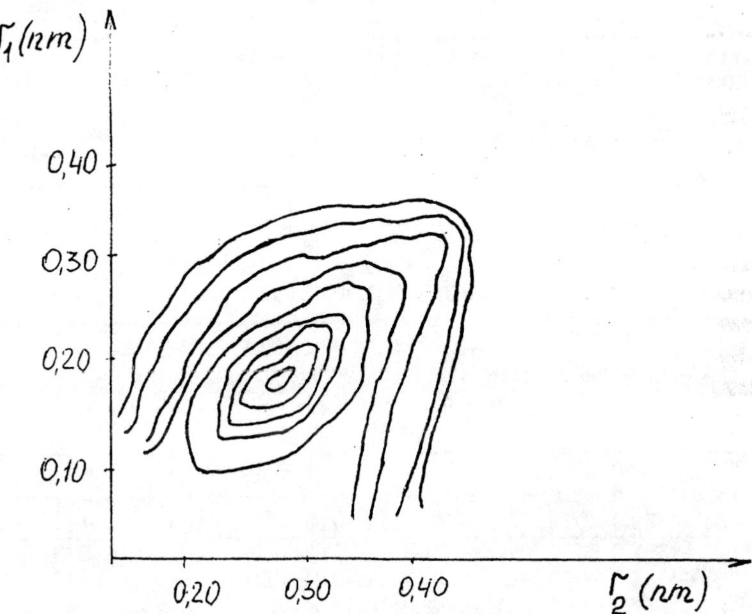

Figure 4.2. Equipotential energetic surface of the reaction $H_{(2)}^+ - H_{(1)}AlCl_3^-$ in the direction 1 (Figure 3.11).

For the model (Figure 3.12), simulating the reaction of cationic center K of complexes with isobutylene, the comparative data on the affinity of H^+, Na^+ and Li^+ to double bond of monomer were obtained (Table 4.8). For all pair interactions there takes place the transfer of positive charge from cation to carbonic atom $C_{(2)}$ of the monomer, which carries sufficient positive charge in differences with other atoms of carbon.

56   V.A. Babkin, G.E. Zaikov, and K.S. Minsker

Atoms $C_{(3)}$ and $C_{(4)}$ are characterized by approximately similar negative charges. Atom $C_{(1)}$ possesses the charge small, for the system H^+-C_4H_8 $C_{(1)}$ atom being charged even positively. This is the indication of the fact, that in the considered model the interaction of cation with double bond does not proceed completely. In particular, it is testified by the equilibrium $C_{(1)}$-$C_{(2)}$ distance equal to 0.146 nm, i.e. shorter than usual σ-bonds of carbon-carbon type equal 0.152 nm. All hydrogen atoms of the monomer are positively charged. The value of dipole moments of the system with the maximum $E_{K^+C_4H_8}$ energy correlates with charge on cation or on the atom of carbon $C_{(2)}$. As it is seen from the Table 4.8 the energies ($\Delta E_{(2)}$), calculated for the interaction of cations H^+ and Li^+ with isobutylene, do not practically differ from each other; for Na it is a little lower.

Table 4.7. Parameters of the interaction of H^+, Na^+, Li^+ cations with $[HAlCl_3]^-$ anion (CNDO/2).

Reaction coordinate, nm	E $K^+H^{(I)}AlCl^3$, kJ/mole	Ebond kJ/mole	Dipole moment	Charges on atoms								$\Delta E_{(2)}$ kJ/mole
			D	Al	Cl$_1$	Cl$_2$	Cl$_3$	H$_1$	H$_2$	Na	Li	
				$H_{(2)}^+[H_{(1)}AlCl_3]^-$								
r$_2$ 0.29	-136700	–	2.3	+0.06	-0.04	-0.02	-0.04	0.00	+0.01	–	–	1920
0.33	-138440	–	2.4	+0.06	-0.03	-0.02	-0.03	-0.01	-0.01	–	–	
0.37	-138250	–	2.5	+0.06	-0.02	-0.02	-0.02	-0.03	-0.03	–	–	
0.39	-138070	–	2.5	+0.06	-0.02	-0.01	-0.02	-0.04	-0.04	–	–	
0.43	-137760	–	2.7	+0.05	-0.01	-0.01	-0.01	-0.05	-0.05	–	–	
r$_3$ 0.13	-0.38	no minimum exists										
				$Na^+[H_{(1)}AlCl_3]^-$								
r$_2$ 0.30	-137590	-3900	3.4	+0.01	+0.02	0.00	+0.02	-0.06	-	+0.01	-	1250
0.35	-137770	-4070	5.9	+0.00	-0.07	-0.03	-0.07	-0.09	-	+0.25	-	
0.38	-137490	-3790	9.9	-0.01	-0.16	-0.06	-0.16	-0.13	-	+0.50	-	
r$_3$ 0.21	-136420	-2730	6.8	+0.00	-0.12	-0.02	+0.09	-0.08	-	+0.13	-	840
0.24	-137220	-3530	9.2	-0.01	-0.14	-0.04	+0.00	-0.09	-	+0.27	-	
0.27	-137330	-3640	12.2	-0.02	-0.16	-0.06	-0.09	-0.10	-	+0.42	-	
0.35	-137240	-3550	14.6	-0.03	-0.18	-0.07	-0.13	-0.11	-	+0.52	-	
				$Li^+[H_{(1)}AlCl_3]^-$								
r$_2$ 0.28	-137050	-3260	4.4	+0.07	-0.11	-0.02	-0.11	-0.09	-	-	+0.19	890
0.32	-137340	-3570	5.7	+0.07	-0.16	-0.05	-0.16	-0.10	-	-	+0.27	
0.34	-137400	-3610	6.5	+0.04	-0.17	-0.06	-0.17	-0.13	-	-	+0.48	
0.36	-137310	-3580	7.7	+0.03	-0.18	-0.07	-0.18	-0.13	-	-	+0.53	
0.45	-137020	-3060	13.0	-0.14	-0.21	-0.09	-0.20	-0.14	-	-	+0.54	
r$_3$ 0.21	-136540	-2750	6.4	+0.01	-0.13	-0.03	-0.02	-0.07	-	-	+0.24	540
0.27	-137050	-3260	13.2	-0.01	-0.18	-0.07	-0.18	-0.10	-	-	+0.55	
0.29	-136960	-3170	14.8	-0.02	-0.19	-0.08	-0.20	-0.10	-	-	+0.59	

Table 4.8. Parameters of interaction of cations with isobutylene (CNDO/2).

Reaction coordinate r_m	$E_{K+C_4H_8}$ kJ/mole	E_{bond} kJ/mole	Dipole moment D	K^+	$C_{(1)}$	$C_{(2)}$	$C_{(3)}$	$C_{(4)}$	ΔE kJ/mole
			H^+–C_4H_8						
r_4	0.10	-88610	-9110	23.60	-	-	-	-	-
	0.12	-89070	-9380	23.50	-	-	-	-	-
	0.14	-89090	-9400	23.60	-	-	-	-	-
	0.16	-89020	-9325	23.80	-	-	-	-	-
	0.18	-88930	-9235	23.90	-	-	-	-	-
r_5	0.13	-89090	-9400	23.60	+0.35	+0.00	+0.26	-0.32	-0.34
	0.14	-89960	-10260	22.30	+0.34	+0.05	+0.18	-0.19	-0.33
	0.15	-89970	-10270	22.30	+0.25	+0.04	+0.20	-0.19	-0.31
	0.15	-89940	-10240	22.30	+0.22	+0.04	+0.20	-0.19	-0.29
	0.15	-89930	-10230	22.20	+0.21	+0.04	+0.21	-0.19	-0.29
			Na^+-C_4H_8						
r_4	0.19	-88400	-9850	19.80	-	-	-	-	-
	0.25	-89070	-10530	17.90	-	-	-	-	-
	0.29	-89090	-10550	16.80	-	-	-	-	-
	0.31	-89030	-10440	16.30	-	-	-	-	-
r_5	0.13	-89090	-10550	16.80	+0.58	-0.18	+0.23	-0.30	-0.29
	0.15	-89710	-11170	14.40	+0.57	-0.18	+0.18	-0.20	-0.30
	0.15	-89690	-11150	14.20	+0.56	-0.19	+0.18	-0.20	-0.30
	0.15	-89660	-11120	13.90	+0.56	-0.20	+0.17	-0.20	-0.29
			Li^+-C_4H_8						
r_4	0.15	-88540	-9910	21.50	-	-	-	-	-
	0.19	-89180	-10540	20.30	-	-	-	-	-
	0.21	-89260	-10630	19.60	-	-	-	-	-
	0.25	-89210	-10560	18.30	-	-	-	-	-
r_5	0.13	-89269	-10630	19.60	+0.38	-0.18	+0.23	-0.29	-0.28
	0.15	-90000	-11370	17.50	+0.38	-0.16	+0.19	-0.18	-0.28
	0.15	-89990	-11360	17.30	+0.37	-0.17	+0.19	-0.18	-0.27
	0.15	-89970	-11340	17.10	+0.37	-0.18	+0.19	-0.18	-0.26

It follows from the comparison of energetic characteristics of the reactions of cation with counterion and isobutylene (see Table 3.7 and 3.8), that except Li^+-isobutylene systems the interaction of cations with anion and isobutylene is simplified at the transition from alkaline metals to hydrogen, which correlates with the well-known data about higher activity of H^+. The preferableness of the attack of anion or isobutylene by cation depends on its nature, however for proton and Na^+ these directions may be considered as proceeding in parallel or as competitive ones.

Conclusions, followed from the calculated data were proved experimentally, studying the systems of $Na^{+\delta}C_2H_5AlCl_3^{-\delta} - HCl(RCl)$-aromatic hydrocarbon (isobutylene).

In dependence on the system composition and conditions there are possible different directions of the reaction with the participation of complexes of alkaline metals:

cationic attack of intraspheric ligands with decomposition of the complex and the initiation of electrophilic processes with the preservation of counterion.

Adding a small amount of HCl (up to Al/H=1:1) into toluene solution of $Na^{+\delta}[C_2H_5AlCl_3]^{+\delta}$ abrupt change of electrical conductivity was observed at the constant ethan extraction.

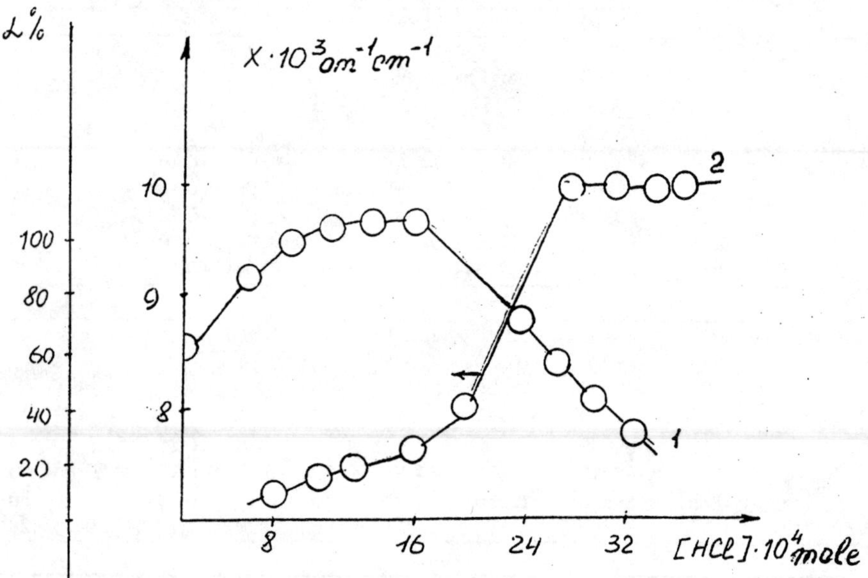

Figure 4.3. The dependence of electric conductivity (1) of $Na^{+\delta}[C_2H_5AlCl_3]^{-\delta}$-HCl-toluene and the amounts (α') of extracted C_2H_6 (2) on admixtures $(Na^{+\delta}[C_2H_5AlCl_3]^{-\delta}]$-2.2·10^{-3} mole, toluene - 1·10^{-5} mole, 298 .

The value of electrical conductivity of the system (æ~10^{-2} Ohm^{-1} cm^{-1}), the one practically equal to the electrical conductivity of Gustavsson complexes [54], points out the formation of protonic complex acid and its stabilization by toluene:

$$Na^{+\delta}[C_2H_5AlCl_3]^{-\delta} \xrightarrow[NaCl]{RCl} H^{+\delta}[C_2H_5AlCl_3]^{-\delta} \xrightarrow{toluene}$$

$$\xrightarrow{toluene} CH_3 - \langle + \bigcirc \rangle - RAlCl_3^-.$$

The precipitation of NaCl residue proves such direction of the reaction, at which metalorganic bond in the anion is destroyed insufficiently (from 7 to 30%, see Figure 4.3).

Separation of the proton reactions with an external acceptor and with $RAlCl_3^-$ anion in the complex acid proceeds at the temperature change. The increase of temperature moves the process practically complete to the side of the reaction of counterion protonation, which is testified by the amount of the extracted ethan, close to the calculated one (Figure 4.4).

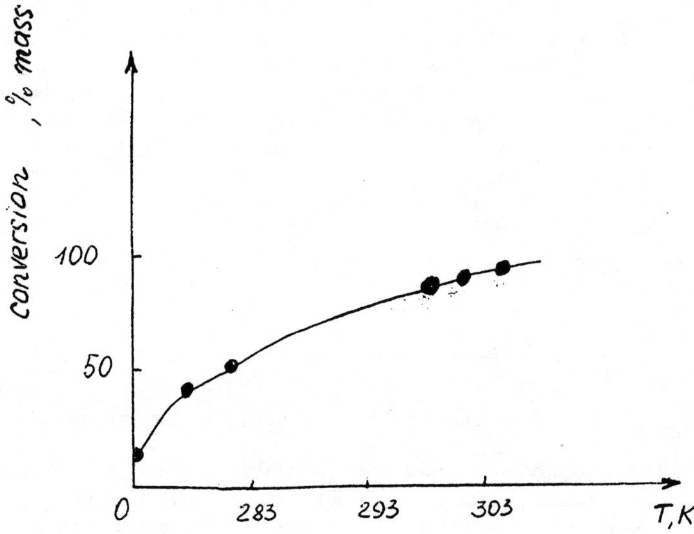

Figure 4.4 The dependence of amount of extracted C_2H_6 on temperature of the reaction: $Na^{+\delta}[C_2H_5AlCl_3]^{-\delta}$-$3.1 \cdot 10^{-3}$ mole, ratio Al:H=1:1, (1 hour).

This reaction decelerates at the temperature decrease. Introduction of isobutylene, which is the acceptor of proton in complex acid, allows to realize its electrophilic properties according to external-spheric reaction of the complex. For example, at the addition of trifluoracetic acid, which displays no properties of the polymerization initiator itself, to $Na^{+\delta}[C_2H_5AlCl_3]^{-\delta}$ in isobutylene at 193-253 K temperatures, the formation of high-molecular products (P=600-2000) is observed.

The study of the systems $Na^{+\delta}[C_2H_5AlCl_3]^{-\delta}$-HCl-isobutylene-aromatic hydrocarbon in the wide range of temperatures from 173 to 193 K gives the possibility to perform the complex of different transformations, in particular: polymerization of isobutylene, alkylation of aromatic hydrocarbon by it, counterion decomposition with elimination of alkan. In other words, protonation of anion and external acceptor in dependence on conditions are competitive reactions.

The substitution of low active cation of alkaline metal (Na^+) by more active R^+ one by adding RCl to the complex and the further polymerization at the addition of isobutylene

$$RCl + Na^{+\delta}[R'\,AlCl_3]^{-\delta} \xrightarrow{-NaCl} R^{+\delta}[r'\,AlCl_3]^{-\delta} \xrightarrow{Mm} RMm^{+\delta}[R'\,AlCl_3]^{-\delta}$$

where Mm - monomer, testifies the electrophilic activity of carbonium anion in relation to the external acceptor. Polymer, being formed at the interaction of the system

$(C_6H_6)_3CCl\text{-}Na^{+\delta}[C_2H_5AlCl_3]^{-\delta}$ with isobutylene contains the end triphenylmethylic groups, which proves the carbcationic activity of the complex forming, as it bases on the data of UV-spectra (absorption in the range of 240-270 nm [107]).

Catalysts of $Na^{+\delta}[C_2H_5AlCl_3]^{-\delta}$ may be applied in chemical and oil industry for utilization of nonstandard fractions of polyisobutylene and other oligo- and polyalkans and alkans. Polymers and oligomers of isobutylene may be transformed into monomer during the process of thermocatalytic degradation under the influence of complex salts – chlorides and alkylchlorides according to the reaction of cationic depolymerization a t 200-400°C. Temperature conditions of the process provide action of the catalyst over the point of phase transition. Monomer output reaches values close to 100%. These catalysts may be applied at the catalyst/carrier ratio of 3:2÷5 (mass parts).

The data on activity of external-spheric cation of the complex were used at working out new complex catalysts for depolymerization of polyisobutylene down to monomer [130].

4.4. Aquacomplexes of Aluminum Chlorides

For aquacomplexes of aluminum chlorides various transformations are characteristic in conditions of the activity display, which depend on nature of aluminum ligands. Besides the models of the initial $R_nAlCl_{3-n}\cdot H_2O$ complexes paper [108] calculated the models, corresponded to transformations of these complexes. For more comfortable consideration the complexes were subdivided into several groups according to chemical structure of Al ligands. The first group consists with aquacomplexes CH_3AlCl_{3-n}, which differ by the ratio of Cl and CH_3 radicals, in particular by $AlCl_3\cdot H_2O$ (I), $CH_3AlCl_2\cdot H_2O$ (II), $(CH_3)_2AlCl\cdot H_2O$ (III) and $(CH_3)_3Al\cdot H_2O$ (IV). Aquacomplexes of the second group differ by one of substitutors a t Al atom in the case that the others are unchanged: $AlCl_3\cdot H_2O$, $CH_3AlCl_2\cdot H_2O$ and $OHAlCl_3\cdot H_2O$.

Characteristic parameters of aquacomplexes of Al compounds with different substitutors are presented in the Table 4.9. The tendency to the formation of aquacomplexes R_nAlCl_{3-n} with H_2O grows in the direction from $(CH_3)_3Al$ to $AlCl_3$, i.e. with the increase of the strength of Lewis acid. This is testified by the character of the change of values of E_o, E_{bond} and $\Delta E_{int}^{H_2O}$. Moreover the comparison of $\Delta E_{int}^{H_2O}$ values for complexes I and II (130 and 100 kJ/mole) and for complexes III and IV (30 kJ/mole) points out the vicinity of the properties of aquacomplexes $AlCl_3\cdot H_2O$ and $CH_3AlCl_3\cdot H_2O$, $(CH_3)_2AlCl\cdot H_2O$ and $(CH_3)_3Al\cdot H_2O$ as Brenstedt acids.

Independent on nature of substitutors at Al atom the aquacomplexes studied are characterized by low Brenstedt acidity in comparison with $R_nAlCl_{3-n}\cdot H_2O$ complexes (high values of $\Delta E_{detach}^{H^+}$ and low values of charges q_{H^+}, Table 4.9), i.e. the proton detachment is nonprofitable. However, it follows from the comparison of ΔE^{H^+} for complexes I - IV, that aquacomplexes I and II are relatively more acidic compounds than III and IV ones. For aluminum derivatives, in which only one substitutor was varied (Table 4.9 I, II and V), the values of energetic parameters of the complex formation ($\Delta E_{int}^{H_2O}$ and E_o) increase in parallel with the increase of Lewis acidity in the following sequence: $CH_3AlCl<ClAlCl_2<OHAlCl_2$. Charges on hydrogen atom

($q_{H+} \sim 0.21$) and the energy of the proton detachment ($\Delta E_{int}^{H^+} = 2000$ kJ/mole) are practically similar for complexes I, II and V, and the values of dipole moments are the higher, the lower the interaction energy $\Delta E_{detach}^{H_2O}$ is (Figure 4.5 and Figure 4.6).

Table 4.9. Characteristic parameters of the model systems $R_nAlCl_{3-n} \cdot H_2O$.

$Z_{XH(I)}$	E_o, kJ/mole	E_{bond}, kJ/mole	Dipole moment D	$\Delta E_{detach}^{H^+}$, kJ/mole	H_2O, ΔE_{int}, kJ/mole
$AlCl_3 \cdot H_2O$ (I)					
0.096	185950	3700	3.50	2000	130
0.11	185960	3720	3.40		
0.12	185880	3620	3.30		
$CH_3AlCl_3 \cdot H_2O$ (II)					
0.096	167220	5860	4.20	1990	100
0.11	167240	5870	4.10		
0.13	167040	5670	3.80		
$(CH_3)_2AlCl \cdot H_2O$ (III)					
0.096	147290	6810	5.50	2400	30
0.11	147330	6740	5.50		
0.13	147880	6760	5.40		
	144880	6050	13.50		
$(CH_3)_3AlCl \cdot H_2O$ (IV)					
0.096	128540	8910	6.90	2400	30
0.11	128550	8940	6.80		
0.13	128490	8860	6.80		
	126100	8150	13.10		
$(OH)AlCl_2 \cdot H_2O$ (V)					
0.096	192870	3620	2.00	2000	150
0.11	192880	3640	1.90		
0.13	192690	3430	1.60		
	190880	3350	7.30		

$\Delta E_{detach}^{H^+}$ – energy of proton detachment from aquacomplex with the calculation of the proton nucleus energy, equal 1600 ± 40 kJ/mole.

$$\Delta E_{int}^{H_2O} = E_o(R_nAlCl_{3-n}) - \left[E_o(R_nAlCl_{3-n}) + E_o(H_2O) \right].$$

The obtained tendency to the increase of the ability of Al compounds of complex formation with H_2O in the sequence $R_3Al \leq R_2AlCl < RAlCl_2 \leq AlCl_3$ correlates with the experimental data. In the mentioned sequence the acidity of aluminum compounds grows as well as their catalytic activity in electrophilic processes [3]. If $AlCl_3$ and $RAlCl_2$, combined in different complexes, possess the properties of the universal acidic catalysts, the systems $R_3Al(R_2AlCl_2)$-H_2O are characterized then by sufficiently weaker acidic properties and cause generally the processes of low-temperature polymerization of strong basic monomers (styrene, isobutylene) [2]. The possibility to form sufficiently stable aquacomplexes (1:1) was displayed experimentally in [39] on the example of $C_2H_5AlCl_2 \cdot H_2O$ which is weak Brenstedt acid.

So, comparing with HCl complexes the ones of $R_nAlCl_{3-n} \cdot H_2O$ type are characterized by low acidic strength, which depends on nature of Al compounds (pKa=+4.4÷+14.4) (Table 4.10). The proton detachment from aquacomplexes is nonprofitable energetically (negative values of ΔE^{H^+}, Table 4.10). Nevertheless high activity is peculiar for them at practically 100-% selectivity of action in the reactions of oligomerization of hydrocarbons (Table 4.11).

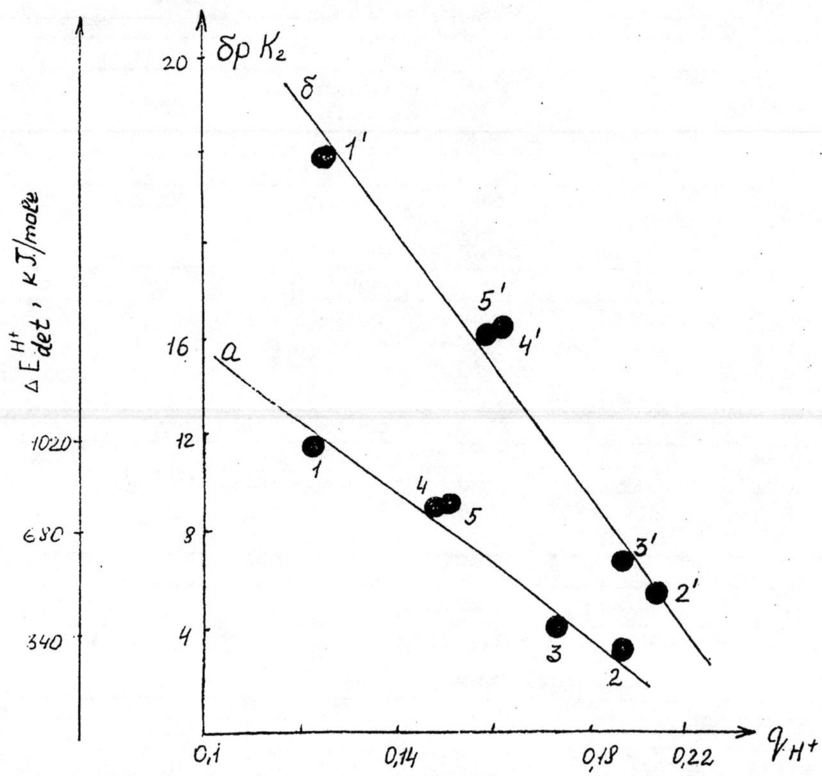

Figure 4.5. The dependence of acidity index (a) and energy of proton detachment (b) on charge on the hydrogen atom: 1,1 - H_2O; 2,2 - $AlCl_3 \cdot H_2O$; 3,3 - $RAlCl_2 \cdot H_2O$; 4,4 - $R_2AlCl \cdot H_2O$; 5,5 - $AlR_3 \cdot H_2O$; R=CH$_3$.

Table 4.10. Calculated parameters of acidic strength of models of aquacomplexes of aluminum halogenides (CNDO/2).

No.	Aquacomplex	$\Delta E_{det\,ach}^{H^+}$	q_{H^+}	δpKa	Group
1.	$AlCl_3 \cdot H_2O$	-390	+0.21	+7.1	
2.	$OHAlCl_2 \cdot H_2O$	-	+0.18	+8.5	I
3.	$(OH)_2AlCl \cdot H_2O$	-	+0.17	+12	
4.	$(OH)_3Al \cdot H_2O$	-	+0.17	+12	
5.	$CH_3AlCl_3 \cdot H_2O$	-420	+0.19	+7.7	
6.	$(CH_3)_2AlCl \cdot H_2O$	-790	+0.16	+14.4	II
7.	$(OH_3)_3Al \cdot H_2O$	-790	+0.16	+14.4	
8.	$EtAlCl_2 \cdot H_2O$	-	+0.20	+7.5	
9.	$(Et)_2AlCl \cdot H_2O$	-	+0.18	+8.5	III
10.	$(Et)_2Al \cdot H_2O$	-	+0.17	+12	
11.	$C_3H_7AlCl_2 \cdot H_2O$	-	+0.19	+7.7	
12.	$(C_3H_7)_2Al \cdot H_2O$	-	+0.17	+12	IV
13.	$(C_3H_7)Al \cdot H_2O$	-	+0.17	+12	
14.	$C_4HgAlCl_2 \cdot H_2O$	-	+0.19	+7.7	
15.	$(C_4Hg)_2AlCl \cdot H_2O$	-	+0.18	+8.5	V
16.	$(C_4Hg)_3Al \cdot H_2O$	-	+0.17	+12	
17.	$OAl_2Cl_4 \cdot H_2O$	-240	-	+4.4	
18.	$CH_3OAl_2Cl_3 \cdot H_2O$	-	+0.21	+7.1	VI
19.	$(CH_3)_2OAl_2Cl \cdot H_2O$	-	+0.21	+7.1	
20.	$(CH_3)_3OAl_2Cl \cdot H_2O$	-	+0.18	+8.5	
21.	$(CH_3)_4OAl_2 \cdot H_2O$	-	+0.18	+8.5	
22.	$HCl \cdot AlCl_3 \cdot H_2O$ [9]	-85	-	+1.5	
23.	$HCl \cdot CH_3AlCl_2 \cdot H_2O$ [9]	-70	-	+1.3	VII
24.	$HCl \cdot OAl_2Cl_4 \cdot H_2O$ [9]	-60	-	+1.0	

Table 4.11. Oligomerization of the mixture of hydrocarbons C_4 in presence of $C_2H_5AlCl_2$* aquacomplexes [109].

No.	Catalyst	Composition of hydrocarbons mixture before the reaction				Conversion i-C_4H_8%	Selectivity n, i-C_4H_8%
		C_3H_6	n, i-C_4H_8	i-C_4H_8	α, β-C_4H_8		
1.	$C_2H_5AlCl_2 \cdot H_2O$	3.0	55.5	37.9	3.5	100	100
2.	$C_2H_5AlCl_2 \cdot H_2O$	3.9	67.5	11.3	17.3	98	100
3.	$C_2H_5AlCl_2 \cdot H_2O$	0.9	45.1	31.1	23.0	99	100
4.	$C_2H_5AlCl_2 \cdot H_2O$**	4.9	63.2	8.5	23.3	100	57

*Conditions: Mixture of C_4-olefins ($3 \cdot 10^{-2}$ l), catalyst - $5.6 \cdot 10^{-2}$ mole/l, dissolvent - heptane, 253 K, $6 \cdot 10^2$ s.

**Dissolvent CH_2Cl_2.

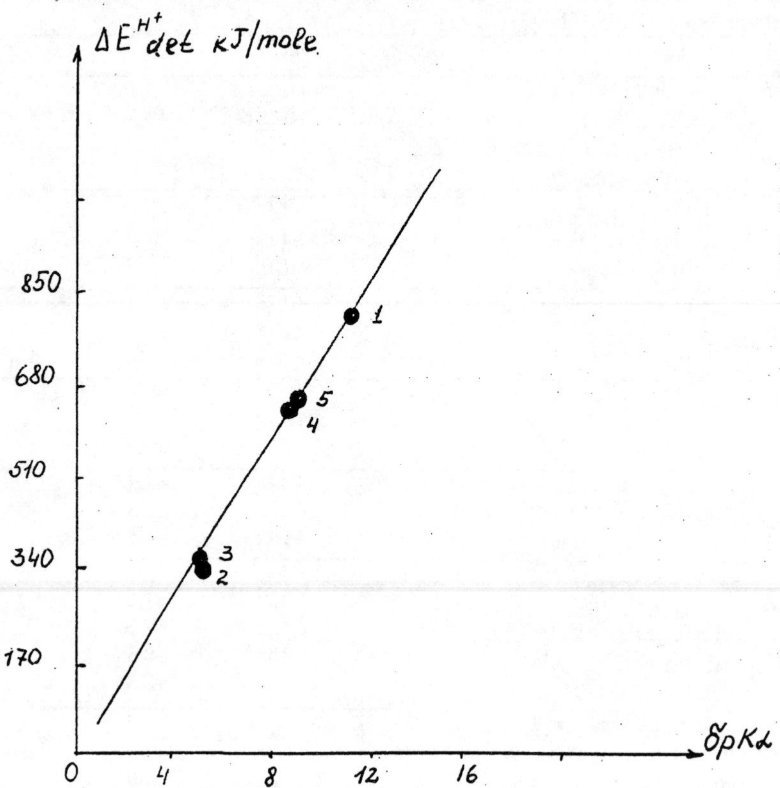

Figure 4.6. The dependence of acidity index on the energy of proton detachment.

Discrepancy of the experimental and calculated data points out the formation of intermediate compounds with the increased acidic strength from aquacomplexes or another mechanism (different from simple acceptance of proton by olefins) of the process excitation. In connection with this circumstance calculated parameters of $R_nAlCl_{3-n}\cdot H_2O$ should be analyzed.

Characteristics of the acidic strength of the models of aquacomplexes R_nAlCl_{3-n} are: q_{H^+} - charge on hydrogen atom of H_2O in the complex, $E^{H^+}_{det\,ach}$ - energy of proton detachment from the complex and pKa - universal index of acidity. They are presented in the Table 4.10. For better suitability aquacomplexes are subdivided into 7 groups, in dependence on nature of radicals of ligand surrounding of Al and probable transformations of aluminum-containing component in R_nAlCl_{3-n} aquacomplexes under the influence of an olefin [17, 19, 49, 110, 113].

Groups I-VI possess general tendency for the decrease of q_{H^+} charge (from +0.2 to +0.16) and the increase of pKa values (from +4.4 to +14.4). It corresponds to the increase of the acidic strength of aquacomplexes of aluminum halogenides in dependence on the amount of atoms, substituted by R radicals, R=OH, CH_3, Et, i-C_3H_7, t-C_4H_9.

In the sequences of corresponding aquacomplexes:

I. $OHAlCl_2 \lozenge H_2O \sim CH_3AlCl_2 \cdot H_2O \sim EtAlCl_2 \cdot H_2O \sim C_3H_7AlCl_2 \cdot H_2O \sim C_4H_9AlCl_2 \cdot H_2O$;

II. $(OH)_2AlCl \cdot H_2O \sim (CH_3)_2AlCl \cdot H_2O \sim (Et)_2AlCl \cdot H_2O \sim (C_3H_7)_2AlCl \cdot H_2O \sim (C_4H_9)_2AlCl \cdot H_2O$;

III. $(OH_3)_3Al \cdot H_2O \sim (CH_3)_3Al \cdot H_2O \sim (Et)_3Al \cdot H_2O \sim (C_3H_7)_3Al \cdot H_2O \sim (C_4H_9)_3Al \cdot H_2O$;

parameters of the acidic strength (q_{H^+} and pKa) testify the fact that the first (I) sequence is characterized by higher acidic strength, comparing with the sequences II and III. This correlates well with the results from the paper [3].

Chemical compositions, collected into the group VII are characterized by the highest acidic strength among the models considered in the Table 4.10. We can specially point out the complexes $HCl \cdot H_2O \cdot AlCl_3$ (pKa~+1.5) and $HCl \cdot H_2O \cdot Al_2Cl_4O_2$ (pKa=+4.4). However, the activity of the initial $R_nAlCl_{3-n} \cdot H_2O$ complex is not connected with intermediate formation of HCl. This fact has found its experimental confirmation at comparative study of selective polymerization of isobutylene [110].

Basing on the experimental and theoretical investigations of aquacomplexes of different aluminum chlorides, the process of AC formation must take into account possible chemical reactions of transformation of complexes under the influence of olefin $>Al-O-Al< \cdot H_2O \cdot HCl$ [110-112]:

$$> Al - O - Al < \cdot H_2O \quad \text{or}$$
$$> C = C < \uparrow \Lambda(\ HCl)$$
$$(-RH) \qquad > C = C <$$
$$R_nAlCl_{3-n} \cdot H_2O + > C = C < \rightarrow R_nAlCl_{3-n} \cdot H_2O \xrightarrow{\quad B \quad} R_nAlCl_{3-n} \cdot H_2O \cdot HCl$$
$$R_nAlCl_{3-n} \cdot H_2O \cdot > C = C <$$

Scheme I

Directions A and B in the scheme I reflect possible transformations of the system component with further complex formation with proton donor, and direction C - the formation of triple complexes with the participation of olefin.

As it is known that under the influence of donors $R_nAlCl_{3-n} \cdot H_2O$ complex transforms into alumoxane in some cases [110-112], then aquacomplex of alumoxane is selected as one of the objects (direction A of the scheme I). Dialkylaluminum chloride, trialkylaluminum give aquacomplexes nonstable even at negative temperatures according to centigrade thermometer, but they form corresponding alumoxanes or their aquacomplexes (for R_2AlCl) [39]. The attention should be paid to the fact that the activity of the systems R_3Al or R_2AlCl and H_2O is usually ascribed to the formation of alumoxane aquacomplexes, which must possess the increased Brenstedt acidity according to [113]. However, the role of alumoxanes in the excitation of polymerization of olefins is set for $(C_2H_5)AlCl-H_2O$ system only [39]. As the hydrolysis of R_nAlCl_{3-n}

in presence of electrodonating compounds is accompanied by the extraction of HCl, which is the acid stronger than water, it is impossible to contradict catalytic action of HCl. This action may be performed through the formation of triple complexes (direction B, scheme I). In this case there are also possible different variants of AC structures (differing by nature of aluminum component for example).

Besides with the types of potential AC pointed out triple complexes $R_nAlCl_{3-n} \cdot H_2O$-olefin are worthy of attention (direction C, scheme I). At the present time participation of olefins on the stage of excitation of electrophilic processes is widely discussed in [54], in particular. For weakly acidic systems of the type of aquacomplexes of aluminum chlorides such approach is of special interest, because the calculation of the features of the catalyst interaction with substrate is the necessary stage in the set of the mechanism of electrophilic reactions.

There exist special experimental facts [39], testifying the fact that electrophilic activity of $C_2H_5AlCl_2 \cdot H_2O$ complex may be connected with the formation of the corresponding complex of alumoxane. It was obtained by IR-spectroscopy method that at the introduction of heptane dissolutions of $C_2H_5AlCl_2$ into toluene, which is close by its basicity to olefins (isobutylene), it is observed the disappearing of characteristic bends of absorption of the initial complex (457, 500 and 605 cm) and absorption bends of alumoxane (485, 575, 626, 1630 and 3200 - 3500 cm). Besides the formation of alumoxane complex at the excitation of oligomerization of difficulty polymerized olefin - disobutylene under the influence of $C_2H_5AlCl_2 \cdot D_2O$, weak absorption bends at 2137 and 2183 $cm^{-1}(V_{C-D})$ were found in the products of the olefin transformation. Probably, this testifies the protonation of olefin by water, presenting in complex with alumoxane:

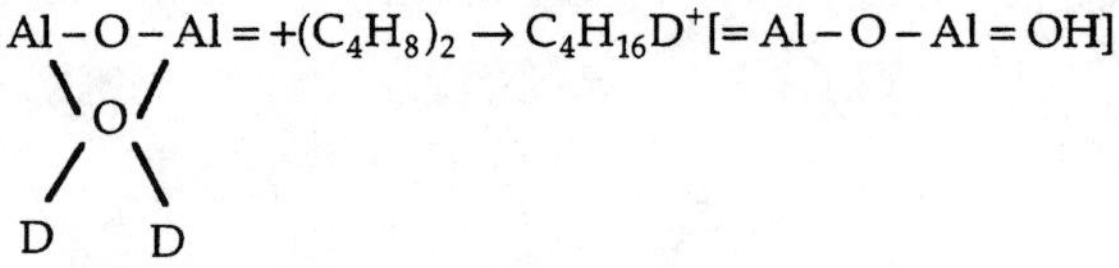

$$Al-O-Al = +(C_4H_8)_2 \rightarrow C_4H_{16}D^+[= Al-O-Al = OH]$$

Scheme II

In this connection the data of quantum-chemical calculations of the model structures of alumoxane aquacomplex are interesting. The initial model is presented on the Figure 3.7. Evidently, atoms of the oxygen $0_{(1)}$ and $0_{(2)}$ are nonequivalent. The calculation of nonequivalency of these atoms was performed according to two methods. In the first case the structure of the alumoxane aquacomplex was taken, in which the bridge was considered equishouldered, and water was coordinated according to Al. As a result of partial optimization it was obtained stable configuration of the complex (see Table 4.12, water disposition is δ, angle $\theta=90°$, $r_y=0.22$ nm, $r_z=0.24$ nm, grouping $Al_{(1)}Cl_{(1)}Cl_{(2)}$ possesses Sp^3-hybridization, and $Al_{(2)}Cl_{(3)}Cl_{(4)}$ - Sp^2-hybridization) (Figure 4.7, structure II).

Table 4.12. Parameters of alumoxane aquacomplex (CNDO/2).

No.	Optimization parameter	E_o, kJ/mole	E_{bond}, kJ/mole	Dipole moment D	$qAl_{(1)}$	$qAl_{(2)}$	$q0_{(1)}$	$q0_{(2)}$
		Planar model						
1	0.21	-283580	-6550	2.48	-0.157	-0.157	+0.053	-0.135
	0.24	-284010	-6970	3.51	-0.095	-0.095	+0.011	-0.159
	0.27	-283940	-6920	3.88	-0.067	-0.067	+0.013	-0.205
2	0.20	-283500	-6470	2.47	-0.156	-0.156	+0.035	-0.120
	0.22	-283590	-6550	2.50	-0.157	-0.157	+0.053	-0.135
	0.23	-283527	-6500	2.48	-0.168	-0.168	+0.067	-0.140
		Distorted model						
3		-284380	-7340	7.37	-0.194	-0.194	+0.001	-0.135
		-284150	-7130	5.29	-0.132	-0.180	+0.014	-0.173
		-284070	-7050	4.07	-0.105	-0.132	+0.015	-0.181
4 Position of								
H_2O	"a"	-284380	-7340	7.37	-0.194	-0.149	+0.001	-0.132
	"b"	-284410	-7390	7.00	-0.200	-0.160	+0.005	-0.167
	"c"	-284400	-7370	6.52	-0.206	-0.140	+0.009	-0.193
		5 Position						
at H_2		-280910	-7130	5.55	-0.247	-0.197	+0.022	-0.326

Continuation of the Table 4.12.

No.	$qCl_{(1)}$	$qCl_{(2)}$	$qCl_{(3)}$	$qCl_{(4)}$	$qH_{(1)}$	$qH_{(2)}$
	Planar model					
1.	+0.062	-0.148	+0.162	-0.148	+0.283	+0.283
	+0.039	-0.106	+0.039	-0.106	+0.236	+0.236
	+0.029	-0.049	+0.029	-0.049	+0.191	+0.191
2.	+0.071	-0.167	+0.071	-0.167	+0.293	+0.293
	+0.068	-0.161	+0.068	-0.161	+0.289	+0.289
	+0.062	-0.148	+0.062	-0.148	+0.283	+0.283
	+0.063	-0.142	+0.063	-0.142	+0.284	+0.284
	Distorted model					
3.	+0.005	-0.081	+0.035	+0.106	+0.197	+0.215
	+0.006	-0.021	+0.031	+0.052	+0.197	+0.215
	+0.007	-0.016	+0.033	-0.030	+0.195	+0.213
4.	+0.005	-0.081	+0.035	+0.106	+0.197	+0.215
	+0.002	-0.023	+0.034	+0.107	+0.202	+0.201
	+0.001	-0.002	+0.039	+0.111	+0.193	+0.189
5.	-0.142	+0.026	-0.031	+0.070	+0.170	-

$$\Delta E_{int}^{H_2O} = E_o(Al_2Cl_4O \cdot H_2O) - \left[E_o(Al_2Cl_4O) + E_o(H_2O)\right]$$

Values of $\Delta E_{detach}^{H^+}$ and $\Delta E_{int}^{H_2O}$ for aquacomplex of alumoxane equal 3240 and 220 kJ/mole, respectively; θ_1, θ_2, θ_3 – values of angles (Figure 4.7).

The calculation has shown that the transition from planar model of symmetric complex (both $AlCl_2$ groups possess Sp^3-hybridization) to complanar one is accompanied by the energy gain of about 380 kJ/mole. The break of equishouldered bridge is energetically profitable. Charges on hydrogen atoms of water are similar (+0.2) and small by their absolute value. The proton detachment from aquacomplex of alumoxane is energetically nonprofitable as judged from the value of energy $\Delta E_{detach}^{H^+} =-$240 kJ/mole.

Further on it was found more stable nonequishouldered structure III ($E_{bond}=-7390$ kJ/mole), the break of the bridge of which is energetically nonprofitable, and both groups possess Sp^3-hybridization (Figure 4.7).

Charges on atoms of hydrogen and oxygen for the structure III equal $q_{H(1)}=+0.21$, $q_{H(2)}$, $q_{0(2)}=-0.21$ and $q_{0(1)}=0$, respectively. Despite high stability of the structure II and nonequivalency of the oxygen atoms, acidity of the alumoxane aquacomplex as judged from the charges on hydrogen atoms of water, did not practically change in comparison with equishouldered model-structure II (Figure 4.7). Symmetric hypothetic structure IIIa, was found energetically less profitable than structure III, and was not considered further on. Structure IV (Figure 4.7) is formed from the structure III resulting the proton detachment. The charge of the system becomes equal -1 (at $H^+\to\infty$, multiplicity remains equal 2S+1=1). Optimization of the structure IV was performed similar to the optimization of the structure III, i.e. first of all it was fixed the disposition of hydrophilic group, and the position of O_1 oxygen, and then the position of OH-group was varied for the obtained minimum of energy. The most stable structure - IV (Sp^3-hybridization), is characterized by the following parameters: $r_3=0.20$ nm, $r_4=0.20$ nm, $r_5=0.19$ nm, $E_{bond}=-7280$ kJ/mole. The comparison of the obtained parameters, correlating with the acidic strength of complexes $\Delta E_{detach}^{H^+}=-420$ kJ/mole, $q_{H^+}=+0.19$ for $RAlCl_2 \cdot H_2O$ and $\Delta E_{detach}^{H^+}=-240$ kJ/mole, $q_{H^+}=+0.21$ for $Cl_4Al_2O \cdot H_2O$ - shows, that whereas the charges on hydrogen atoms of both complexes are close, the proton detachment is made simple in the case of alumoxane aquacomplex. However, judging from the value of pKa=+4.4, it also must display low catalytic activity.

Another possible way of the AC formation (direction B of the scheme 1) is connected with the appearance of chlordinated hydrogen in the system and the formation of $R_nAlCl_{3-n} \cdot H_2O \cdot HCl$ complexes (Figure 4.9). The values of $\Delta E_{detach}^{H^+}$ in the minimum of energy for the main state for triple complexes $AlCl_3 \cdot H_2O \cdot HCl$, $RAlCl_2 \cdot H_2O \cdot HCl$, $Al_2Cl_4O \cdot H_2O \cdot HCl$ equal -85, -70 and -60 kJ/mole, respectively. The comparison of these values with the corresponding values for double complexes (Table 4.13) shows, that the proton detachment is made simple by 3-5 times in the case of the formation of triple complexes. However negative values of $\Delta E_{detach}^{H^+}$ point out the fact that spontaneous ionization, characteristic for complexes $AlCl_3 \cdot HCl$ [19], takes no place in the present case. Consequently, the activity of the initial R_nAlCl_{3-n} aquacomplex is not connected with intermediate formation of hydrogen chloride. This important notion is proved experimentally at the comparative study of selective polymerization in presence of complex catalysts $C_2H_5AlCl_2 \cdot H_2O$, $C_2H_5AlCl_2$ and $C_2H_5AlCl_2 \cdot H_2O \cdot HCl$.

In the case of addition of HCl to $C_2H_5AlCl_2 \cdot H_2O$ selectivity of the process decreases (Figure 4.8) and approaches to the selectivity of the process, initiated by $C_2H_5AlCl_2 \cdot HCl$ complexes. This testifies the fact, that the activity of

$C_2H_5AlCl_2 \cdot H_2O$ in selective polymerization of olefins is not connected with the formation of triple complexes $C_2H_5AlCl_2 \cdot H_2O \cdot HCl$. Thus, the transformation of $RAlCl_2 \cdot H_2O$ into alumoxane and probable formation of triple complexes does not correspond with probable AC (direction B of the scheme 1). Consequently, pointed out transformations (directions A and B of the scheme II) are not the principal ones from the point of view of the initiation of cationic polymerization of olefins. The process of polymerization was controlled according to the data of chromatographic analysis of the reactionary mass, selectivity of polymerization of isobutylene from olefins mixture in parts of unit were determined as the correlation of the transformation degree of isobutylene to total transformation of all olefins with the accuracy of ±5%.

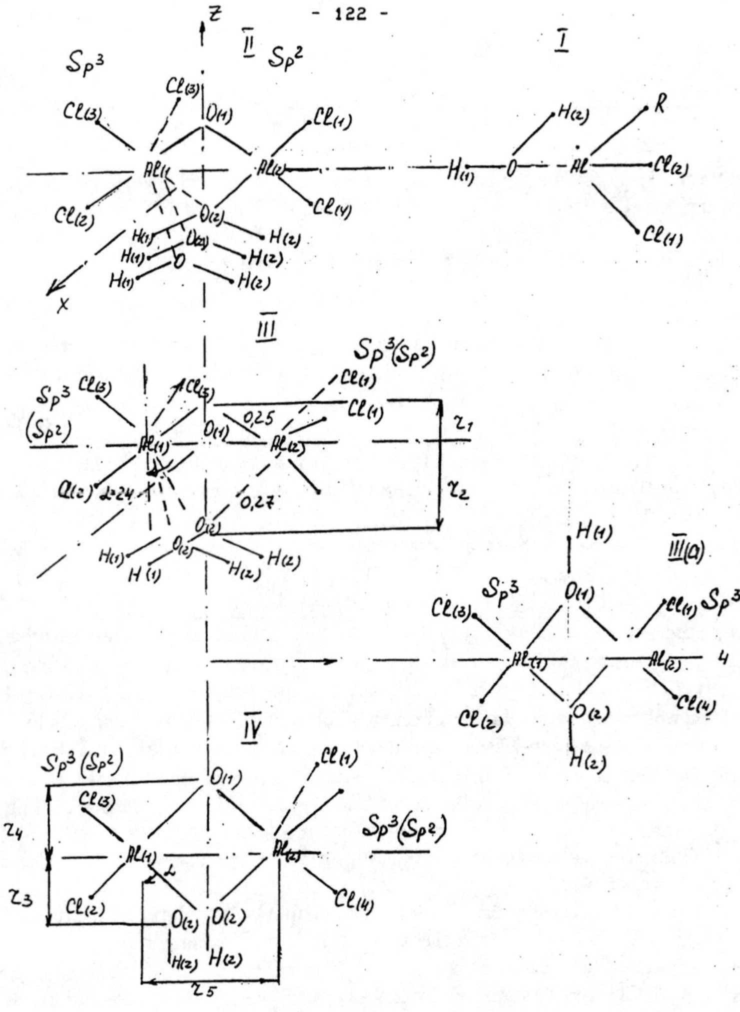

Figure 4.7. Model structures of alumoxane aquacomplexes (CNDO/2).

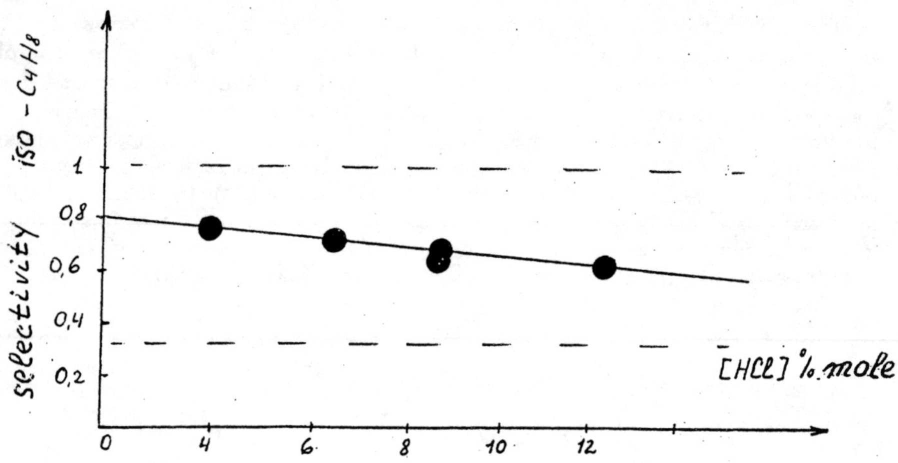

Figure 4.8. The dependence of selectivity of polymerization of isobutylene on admixtures of HCl to C2H5AlCl2◊H2O (1); selectivity of the process in presence of C2H5AlCl2◊HCl (2); in presence of C2H5AlCl2◊HCl - (3); 250 K. Polymerization conditions: [C4] - 3◊10-6 mole◊g-1, [C2H5AlCl2]=8◊10-3 mole/l.

Aquacomplexes of aluminium halogenides, corresponded to the groups I-VI, are characterized, judged to the values of q_{H+} and pKa by low Brenstedt acidity. The highest acidic strength among the compounds of the I-II groups of complexes was obtained for $Al_2Cl_4O \cdot H_2O$ (pKa~+4.4), $AlCl_3 \cdot H_2O$ (pKa~+7.1) and $AlCl_2Et \cdot H_2O$ (pKa~+7.5), which cannot be connsidered as universal acidic catalysts for the most of electrophilic reactions in comparison with the systems $R_3Al(R_2AlCl) \cdot H_2O$, characterized by weaker acidic properties [3]. In this connection the specific interest is presented by the calculation of interaction of aquacomplexes - $H_2O \cdot AlR_nCl_{3-n}$, in particular by the interaction of $AlCl_4O \cdot H_2O$, $AlCl_3 \cdot H_2O$ and $AlCl_2Et \cdot H_2O$ with olefins (ethylene, propylene, isobutylene). Moreover in the number of cases the electrophilic activity of $AlCl_2Et \cdot H_2O$ complex is connected with the formation of the corresponding aquacomplex of alumoxane. Considering the dynamics of the process in conditions of gradual adding of the monomer to the solution of $AlCl_2Et \cdot H_2O$ at -30÷-78°C it was obtained that the reaction of alumoxane formation proceeds by 7-8%, i.e. the reactionary system possesses simultaneously both the initial $AlCl_2Et \cdot H_2O$ system and the product of its transformation - aquacomplex of alumoxane [108, 114]. This circumstance makes the determination of the real nature of AC difficult. The analysis of its quantum-chemical parameters of aquacomplexes interaction with olefins will allow to state the real structure of AC, responsible for the excitation of the process of polymerization of olefins.

Preliminarily it was investigated the influence of olefin (isobutylene) on the behavior of $AlCl_3 \cdot H_2O$ and $OHAlCl_2 \cdot H_2O$ complexes, corresponded to the direction C in the scheme I. It was determined the most energetically profitable direction of the attack of $R_nAlCl_{3-n} \cdot H_2O$ complex by isobutyene by means of displacement of olefin in the plans XOY, ZOX, ZOY. This disposition of isobutylene is presented on the Figure

4.10, and the parameters, characteristic for this coordination, are presented in the Table 4.13.

Table 4.13. Parameters of model systems $H_2O \cdot R_nAlCl_{3-n}$ - isobutylene (CNDO/2).

System	E_o, kJ/mole	E_{bond}, kJ/mole	$q_{C(1)}$	$q_{C(2)}$	q_{10}	ΔE_{int}^m kJ/mole	$\Delta E_{int}^{H_2O}$ kJ/mole	$\Delta E_{theor}^{H^+}$
$AlCl_3 \cdot H_2O \cdot C_4H_8$	-273640	-14360	+0.24	-0.20	-0.25	280	405	1625
$CH_3AlCl_2 \cdot H_2O \cdot C_4H_8$	-134990	-16450	+0.23	-0.23	-0.25	290	420	1620
$OHAlCl_2 \cdot H_2O \cdot C_4H_8$	-280540	-14350	+0.22	-0.19	-0.36	280	390	1615

ΔE_{int}^m - energy of interaction of isobutylene with aquacomplex, $r_x=0.25$ n, see Figure 4.10.

Comparison of $\Delta E_{int}^{H_2O}$ and $\Delta E_{detach}^{H^+}$ values for complexes $OHAlCl_2 \cdot H_2O$, $AlCl_3 \cdot H_2O$, $RAlCl_2 \cdot H_2O$ in presence (Table 4.13) and in absence (Table 4.10) of isobutylene allows to conclude that at coordination of isobutylene the bond energy of $R_nAlCl_2 \cdot H_2O$ increases by 3-4 times, and Al-O bond energy decreases approximately by 20%. These results correlate with the scheme of H-initiation of the process of polymerization of olefins in presence of $R_nAlCl_2 \cdot H_2O$ complexes. The attention should be paid to the fact, that for different complexes similar vlaues of ΔE_{int}^m were obtained (in the ranges of the CNDO/2 method accuracy). Other characteristics of triple complexes depend weakly on nature of the aluminium component. That is why as the initial aquacomplexes $R_nAlCl_2 \cdot H_2O$ and alumoxane aquacomplex possess similar values of q_{H^+}, $\Delta E_{theor}^{H^+}$ and pKa. Their behavior at the interaction with olefin will be similar.

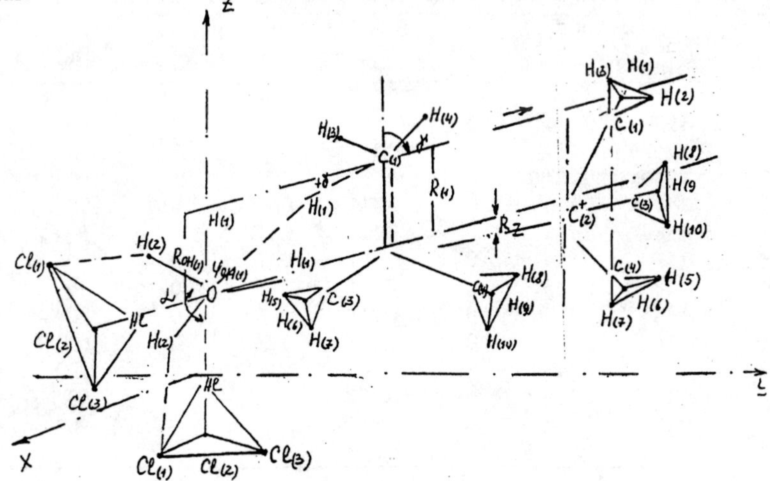

Figure 4.9. The model of interaction of aquacomplex with olefin.

For more detailed investigation of the role of olefin in the process of AC formation the dynamics of interaction of isobutylene with AC may be considered as proceeding with every aquacomplex, with $AlCl_3 \cdot H_2O$ for example. Optimization of $AlCl_3 \cdot H_2O$ - isobutylene complex structure was performed by six parameters, playing the ascertain role in the reaction: length of >C=C< double bond ($R_{(2)}$), length of $O-H_{(1)}$ bond, angle of $AlCl_3 \cdot OH_{(2)}$ fragment deflection in ZOY plane in relation to isobutylene (α), angle of the attack of olefin by proton ($\varphi_{OH_{(1)}}$), displacement of $AlCl_3 \cdot H_2O$ molecule according to $Z_{(R_z)}$ axis and angle of forming tetrahedron (γ) (Figure 4.9). Distance $R_{OC+_{(2)}}$ was selected as the reaction coordinate.

Variable parameters and the change of E_0 value of $AlCl_3 \cdot H_2O$-isobutylene system are presented in the Table 4.14.

It should be mentioned that the real number of variable geometrical parameters is more than six. The stage of olefin coordination includes 4 steps. Initial valu of $R_{OC_{(2)}}$ was accepted equal 0.300 nm, and the energy of the first step - zero. Characteristic features of the coordination stage are rotation of $AlCl_3 \cdot OH_{(2)}$ fragment the complex molecule by angle a=90° (it does not change on the further stages) and orientation of $O-H_{(1)}$ bond to the side of isobutylene. Stages 2 and 3 possess the decrease of E_0 value down to -20÷-40 kJ/mole, and the bond length $R_{OH_{(1)}}$ increases up to 0.108 and 0.111 nm, respectively. The next stage possesses the increase of E_0 value of the system, parameters $R_{(1)}$, $R_{OH_{(1)}}$ and γ being constant and values R_z and $\varphi_{OH_{(1)}}$ decreasing. Thus the stage of coordination of the interaction of isobutylene with aquacomplex is characterized by the energy minimum (Figure 4.10).

Table 4.14. Changing variable parameters and E_0 at the interaction of $AlCl_3 \cdot H_2O$ aquacomplex with isobutylene (CNDO/2).

Step No.	$R_{OC}^{(2)}$ nm	$R_{(1)}$ nm	γ grad	$R_{OH_{(1)}}$ nm	R_z nm	$\varphi_{OH_{(1)}}$ grad	α grad	$E_0,$ kJ/mole
1.	0.300	0.134	0	0.105	0.40	70	0	0
2.	0.280	0.134	0	0.108	0.40	70	90	-20
3.	0.260	0.134	0	0.111	0.40	70	90	-40
4.	0.240	0.134	0	0.111	0.40	60	90	-35
5.	0.220	0.136	0	0.111	0.20	40	90	-5
6.	0.180	0.136	0	0.111	0.20	0	90	0
7.	0.170	0.136	0	0.111	0.20	0	90	+5
8.	0.165	0.136	0	0.111	0.20	0	90	+5
9.	0.163	0.136	0	0.111	0.20	0	90	+10
10.	0.160	0.146	0	0.158	0.20	20	90	-420
11.	0.155	0.148	7	–	0.20	–	90	-480
12.	0.150	0.148	19	0.177	0.20	–	90	-510
13.	0.142	0.148	–	0.212	0.10	–	90	-620
14.	0.139	0.148	–	0.200	0.10	–	90	-610

The second stage of the process of initiation of isobutylene polymerization starts with the break of $C_{(1)}=C_{(2)}$ π-bond of the olefin (Table 4.14, step 5). E_0 reaches the minimum at the value of 10 kJ/mole. Consequently, energetic barrier is 50 kJ/mole. At $R_{(1)}$0.160 nm and the attack angle $\varphi_{OH_{(1)}}$=20° π-bond length increases from 0.136 to

0.146 nm, and O-H$_{(1)}$ bond does practically break (Table 4.14, stage 10), parameters R$_{(1)}$, R$_{OH(1)}$ and γ become independent. Particle H$_{(1)}^{+\delta}$ attacks mostly hydrogenated C$_{(1)}$ atom, forming tetrahedron C$_{(2)}$H$_{(1)}$H$_{(3)}$H$_{(4)}$ (γ=19°). On the step 11 elongation of C$_1$=C$_2$ bond is observed up to 0.148 nm and the decreae of E$_0$ value down to -480 kJ/mole also (Figure 4.10). The stage of carbcation formation completes the attachment of H$^{+\delta}$ particle to C$_{(1)}$ atom, C$_{(2)}^{+}$ carbcation in Sp^2-hybridization being formed simultaneously. Atoms C$_{(1)}$C$_{(2)}$C$_{(3)}$C$_{(4)}$ fall within a single plane, but the interaction with negatively charged [AlCl$_3$·OH$_{(2)}$]$^{-\delta}$ fragment of the complex causes their coming out from it forming tetrahedral structure (12-14 steps). This process is accompanied by additional energy gain of 580 kJ/mole.

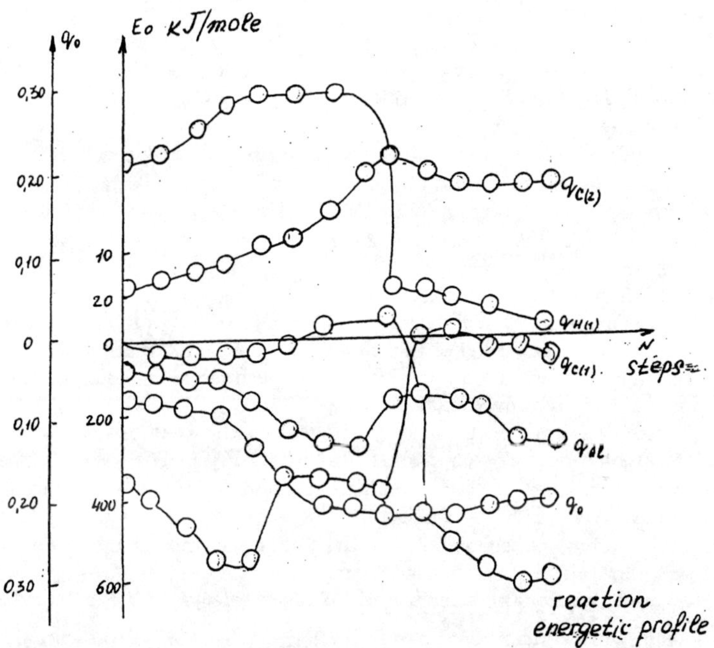

Figure 4.10. The change of total energy of the system and charges on the atoms at the interaction with isobutylene: a) the stage of olefin coordination; b) the stage of π-bond break; c) the stage of carbcation formation.

The change of charges on atoms of AlCl$_3$·H$_2$O complex at the interaction with isobutylene are presented on the Figure 4.10. As isobutylene approaches active O-H$_{(1)}$ bond, the charge on H$_{(1)}$ atom increases from +0.22 to +0.29, which according to correlation dependence pKa-q$_{H+(1)}$ relates to the increase of pKa values from 0 to -15. The detachment of H$_{(1)}^{+\delta}$ (decreasing part of the energetic curve, Figure 4.10) takes place at the acidic strength value a little lower (q$_{H+}\sim$+0.2; δpKa=-14). Positive charge on a-carbonic atom $C_{(2)}$ increases gradually from +0.06 to +0.2 at the monomer approaching aquacomplex by means of displacement of π-electon density of double bond to $C_{(1)}$ atom. Oxygen atom of H_2O does not endure sufficient changes on the whole (Figure 4.10), whereas it receives additional electron density on the stage of monomer coordination. Probably, this is connected with partial charge transfer from the oxygen atom to Al atom on the stage, preceding the break of π-bond.

Approximate equality of values of positive charge on the carbon atom $C_{(2)}$ and negative one on the oxygen atom (taking into account its transfer to Al atom) testifies the formation of AC as polarized intermediate $[Cl_3AlOH]^{-\delta}$... $(CH_3)_3C^{+\delta}$. Similar structure was obtained by calculations for complex with alcohol also, performing carbcationic activity [19]. Estimation of the bond orders of $O\text{-}H_{(2)}$ (0.9) and $O\text{-}C_{(2)}$ (0.8) points out that the introduction of the next monomeric unit will be performed preferably by the bond --- $O^{-\delta}....C_{(2)}^{+\delta}$.

Thus, the coordination of monomer with catalyst promotes the display of strong Brenstedt acidity by aquacomplexes of chlorides and the formation of AC, weakening (break) of $O\text{-}H_{(1)}$ bonds of the catalyst and $C_{(1)}{=}C_{(2)}$ bonds of olefin proceeding simultaneously.

Preliminary complex formation of H_2O with $AlCl_3$ is concluded not only in its partial excitation and provision of the direction of the bond attack by monomer, but in "rigid" linking of $O\text{-}H_{(2)}$ group with $AlCl_3$. There exist different viewpoints on the role of counterion in the interaction of carbcation with olefin. Depending on nature of catalytic system and other conditions, counterion may make the attack of carbcation by monomer difficult [115] as well as it may promote olefin activation (the case of negative electrophilic polymerization) [5] or partcipate equitably with carbcation in the disclosure of double bond of the monomer ("push-pull mechanism") [5]. However, these assumptions are based on indirect experimental results, so they cannot be considered as strictly proved ones.

The role of counterion is mostly sufficient in the case of weakly acidic complex catalysts, which are aquacomplexes $R_nAlCl_{3-n}\cdot H_2O$. Theoretically set connection between the processes of carbcation and counterion formation at the interaction of aquacomplex $AlCl_3\cdot H_2O$ with isobutylene supposes the behavior of carbcationic centers, depending on counterion on the stage of the chain growth of cationic polymerization also.

The most reactionary structure of AC was selected for the calculation, which corresponds with contact ionic pair $(CH_3)_3^{C_{(2)}^{+\delta}}...(Cl_3AlOH)^{-\delta}$ possessing the distance between cation and anion of 0.142-0.150 nm (Figure 4.9).

Ethylene was selected as the monomer.

Taking into account the character of the reaction of monomer and counterion with carbcationic center, transitional state may be presented as the reactions as nucleophilic substitution:

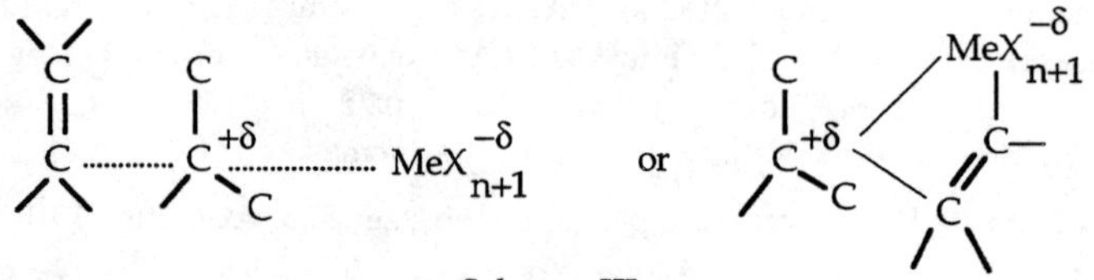

Scheme III

Contrary to B structure, A structure must provide more beneficial attack of carbcation by ethylene because of the absence of shielding by counterion. On the other side, steric factors play sufficient role in the transition state of the reactions proceeding by S_{N2} mechanism. The presence of steric substitutors in carbcationic center (the case of growing carbonium ions) makes the approach of olefin difficult, and creates the state A less profitable, than B one.

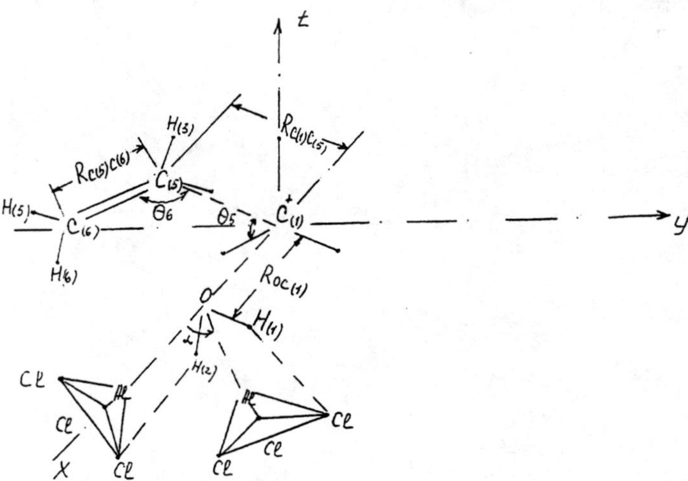

Figure 4.11a. The attack of cationic pair by olefin according to the scheme 3(A).

Figure 4.11b. The attack of cationic pair by olefin according to the scheme 3(B).

There were calculated the models, related to the interaction of ethylene with ionic pair according to the Scheme III, in order to set the most probable direction (A or B) of the reaction.

Figure 4.11 shows the case of the attack of contact ionic pair by ethylene from the side opposite to the disposition of anionic fragment. The following six parameters were selected for optimization of ethylene disposition in relation to carbcation: $R_{C_{(6)}C_{(5)}}$ – length of $C_{(6)}C_{(5)}$ double bond, $\theta_{(1)}$ – monomer deflection angle in the plane ZOX, $R_{OC_{(1)}}$ – distance from carbcation to counterion, $\theta_{(2)}$ and $\varphi_{(2)}$ – angles of deflection of atoms $H_{(2)}$, $H_{(3)}$, $H_{(4)}$ and $H_{(5)}$ in the planes XOY and ZOX, respectively.

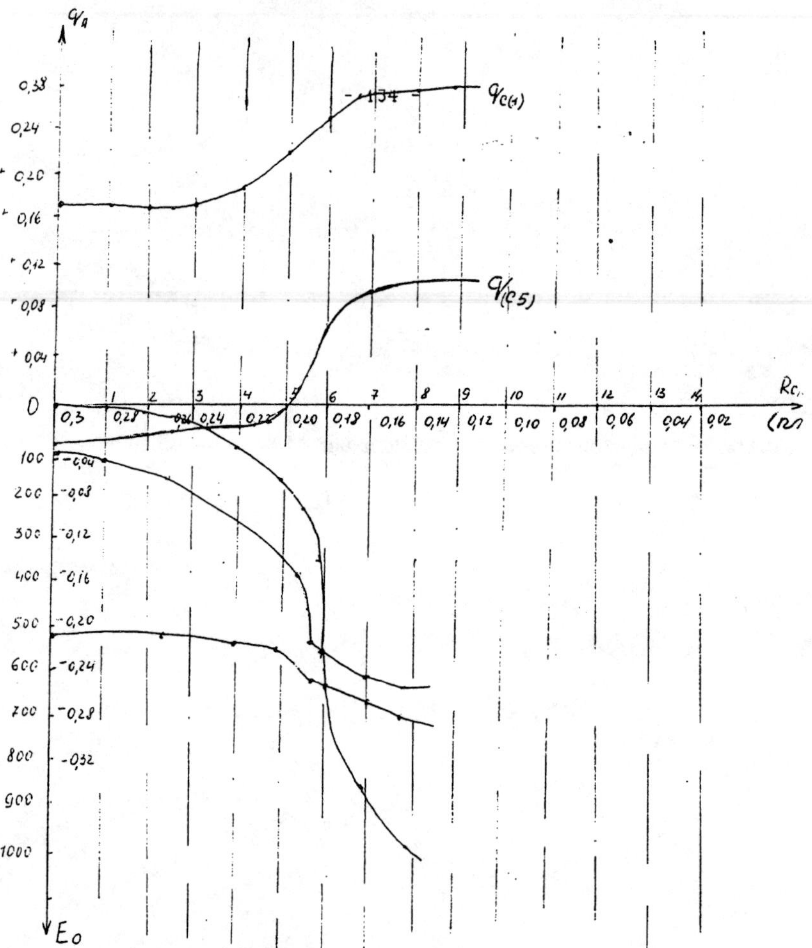

Figure 4.12. The change of charges and total energy at the attack of ionic pair by ethylene according to the scheme 3(A) (CNDO/2).

Figure 4.12 shows the changes of charges on atoms of the reactionary center and values of E_O according to the reaction coordinate, the role of which is played by the distance $R_{C_{(1)}C_{(5)}}$.

Positive charge on carbon atom $C_{(1)}$ of carbcation and negative charge on the oxygen atom increases. Carbon atoms in ethylene molecule also change their charges, the growth of negative charge on $C_{(6)}$ atom being more noticeable, than that of positive one on $C_{(5)}$ atom. Nevertheless, double bond $C_{(5)}=C_{(6)}$ endures no sufficient changes ($P_{C_{(5)}C_{(6)}}=1.8 \div 1.9$, $R_{C_{(5)}C_{(6)}}=0.135$ nm). The distance $R_{OC_{(1)}}$ is practically constant at the decrease of $R_{C_{(1)}C_{(5)}}$ from 0.30 to 0.148 nm. Absence of loosening of olefin double bond testifies the fact that the attack of contact ionic pair by ethylene is difficult from the side opposite to anion. Low probability of the reaction proceeding according to the Scheme IIIa is in contradiction with the conclusions of theoretical work [116]. According to the work mentioned this reaction probably models the reaction of the chain growth. It is imagined that the main argument – nonbarrier character of the process of olefin coordination with carbcationic center of ionic pair – cannot be the basis for the statement that the whole reaction possesses quantum (tunneling) character. This fact is confirmed by the data from the paper [116], which points out adverse charges separation on the stage of the olefin π-bond break.

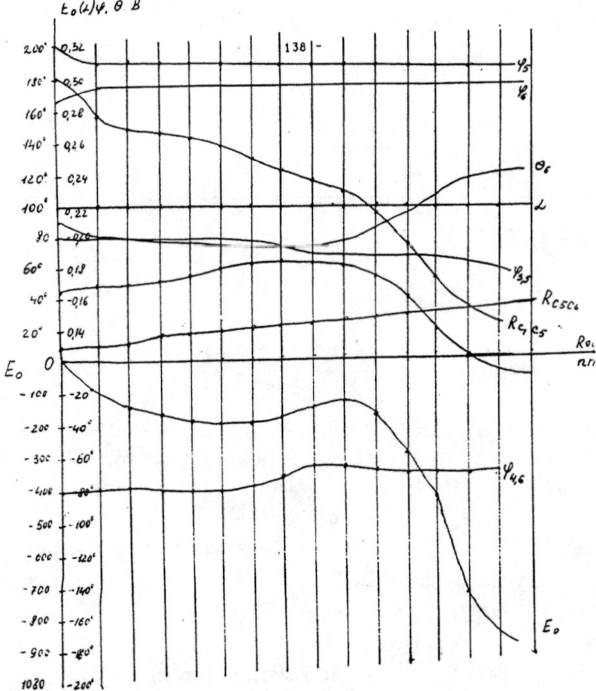

Figure 4.13. Variable parameters and total energy of the system for the reaction of ethylene with contact ionic pair according to the scheme 3(B) (CNDO/2).

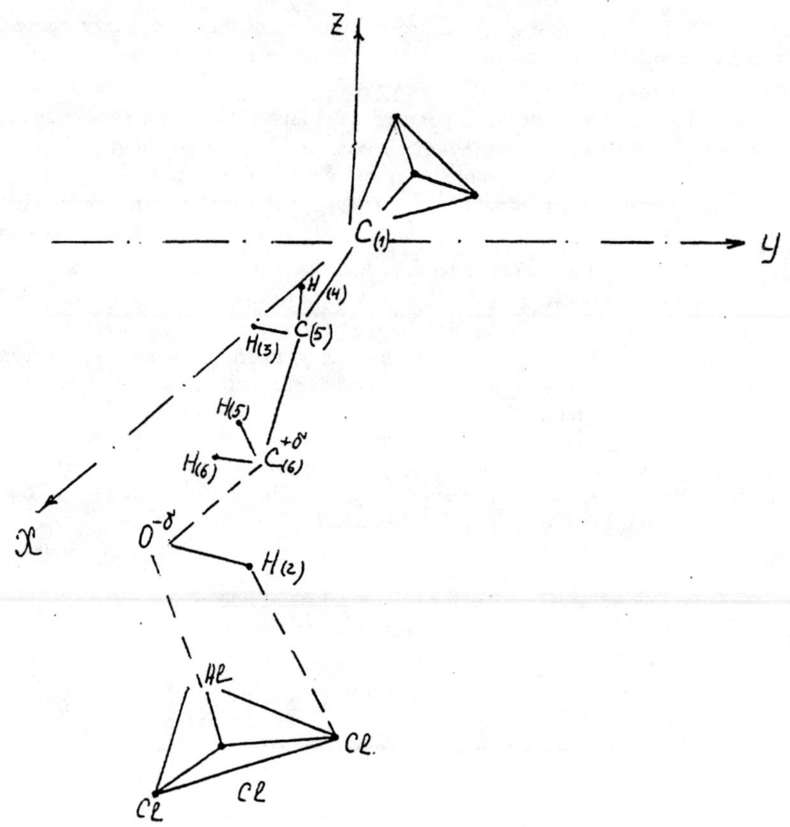

Figure 4.14. The end structure after olefin introduction.

Considering the model, relating to the Scheme IIIb, optimization of the disposition of ethylene molecule according to ionic pair is performed in the following way. First of all the most stable configuration of the system was determined at ethylene rotation round the bond O-$C_{(1)}$ (rotation radii $R_{C_{(1)}C_{(5)}}$ change in the range from 0.29 to 0.50 nm), $R_{C_{(1)}C_{(5)}}$=0.30 nm (Figure 4.11b). E_0 value was set equal zero. Further on the angles of ethylene deflection were changed in the planes ZOX ($\theta_{(5)}\theta_{(6)}$) and ZOY ($\varphi_{(5)}\varphi_{(6)}$), length of double bond $R_{C_{(5)}C_{(6)}}$, rotation angles of atoms $H_{(3)}$, $H_{(4)}$, $H_{(5)}$, $H_{(6)}$ in the planes ZOX ($\varphi_{(3-6)}$), deflection angle of counterion in the plane ZOX (α) and legnth of the forming bond $C_{(1)}C_{(5)}$ ($R_{C_{(1)}C_{(5)}}$). The distance from carbcation to counterion $R_{OC_{(1)}}$ was chosen as the reaction coordinate.

The change of parameters and values of E_0 during the reaction are shown on the Figure 4.13.

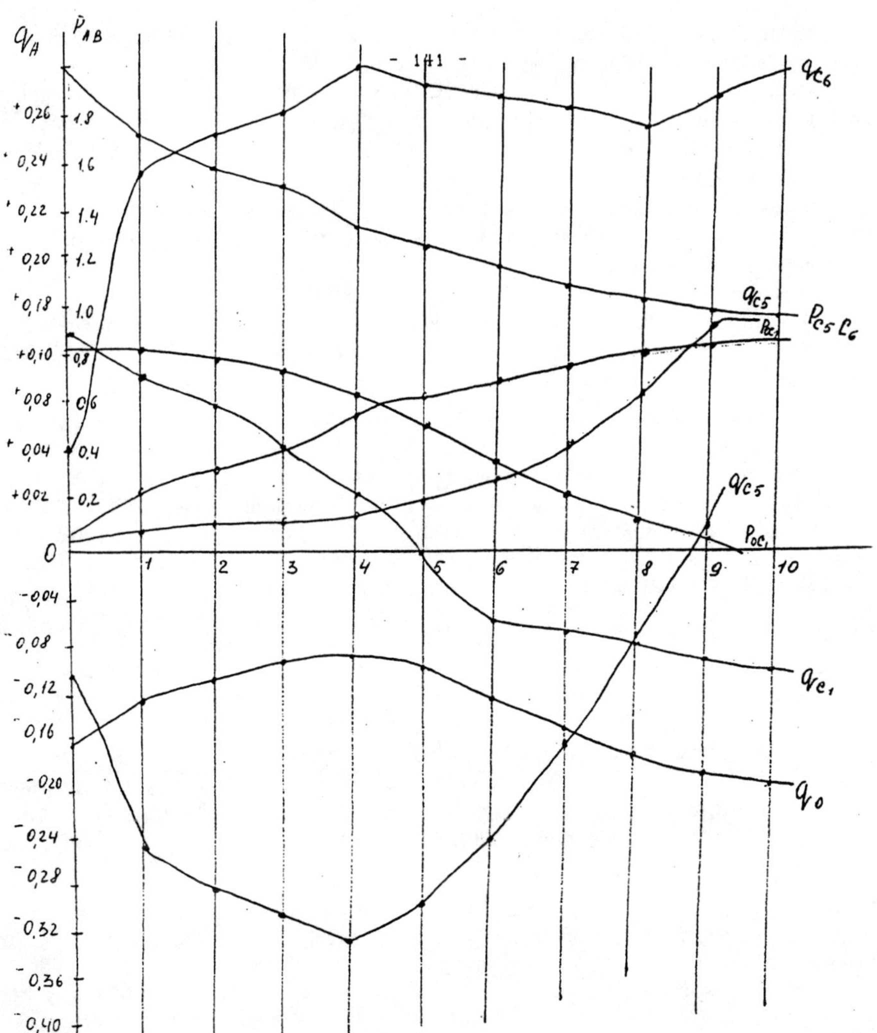

Figure 4.15. The change of charges of atoms and bonds order at the attack of contact ionic pair by olefin from the side of counterion (CNDO/2).

Coordination stage ($R_{OC_{(1)}}$ is varied in the range from 0.142 to 0.160 nm) is characterized by counterion $[AlCl_3 \cdot OH]^{-\delta}$ deflection on the angle $\alpha = 100°$ (which does not change during the further stages) and by the decrease of E_0 down -175 kJ/mole.

At the second stage ($R_{OC_{(1)}} = 0.160 \div 0.180$ nm) energy of the system increases up to -135 kJ/mole. Consequently, energetic barrier is 40 kJ/mole.

The break of $C_{(5)}C_{(6)}$ bond is finished on the stage of the end structure formation, and the introduction of ethylene by $O\text{-}C_{(1)}$ bond takes place (Figure 4.14).

The analysis shows that positive charge on carbon atom of carbcation $C_{(1)}$ decreases, and negative one on olefin atom increases (Figure 4.15).

Deficit of electrons on α-carbonic atom $C_{(6)}$ increases also by means of displacement of π-electron of the double bond by direction to $C_{(5)}$ atom. Total negative charge on counterion on the whole endure no sufficient changes, whereas it displays the tendency to the decrease. The order of the double bond $C_{(5)}=C_{(6)}$ decreases from 2 to 1. $O-C_{(1)}$ bond is weakening simultaneously down to the brea, and the bonds $C_{(6)}-O$ and $C_{(1)}-C_{(5)}$ are formed. Small change of positive charge on $C_{(1)}$ atom and simultaneous sufficient weakening of the double bond $C_{(5)}=C_{(6)}$ testifies the sufficient role of overlapping of p-orbitals of carbcation and olefin on this stage of interaction.

Evidently, interaction of the next molecule of ethylene with carbcation will be performed by the new formed ionic bond, because its parameters ($R_{C_{(6)}-O}$-0.142 nm, $P_{C_{(6)}-O}$=0.89) are close to these of the initial $C_{(1)}-O$ bond of the contact ionic pair. The condition of the reaction proceeding is planar disposition of atoms $C_{(6)}$, $C_{(5)}$, $H_{(5)}$ and $H_{(6)}$.

The calculation performed points out the important role of counterion in the process of interaction of olefin with carbcationic center according to the Scheme IIIb. Anionic fragment of cationic pair participates directly in electron transitions, connected with disclosure of the double bond of olefin, whereas on the whole it endures no sufficient changes. It should be expected, because the system returns into the electron stage, similar to the initial one, at the end of the reaction. The change of charges on carbon atoms of olefin and carbcation testifies the coordinated character of interaction of carbcation with olefin. Really, the weakening of carbcation-counterion bond (dissociation of the contact ionic pair) is not accompanied by the expected increase of charges on components of the ionic pair and E_0 increase. Adverse separation of charges is compensated by olefin coordination in relation to active $C_{(1)}-O$ bond. The process is accompanied by the disposition of counterion in space, that it creates the minimum obstacles for the attack of AC by ethylene. It is interesting that if the formation of new $C_{(6)}^{+\delta}$-anion bond is strictly corresponded to the break of the initial $C_{(1)}^{+\delta}$-anion bond, then the formation of $C_{(1)}^{+\delta}$-monomer bond is a little delay in relation to the break of aliqout bond. According to small variations of the charge on carbon atom of carbcation this testifies the sufficient role of overlapping of p-orbitals on the present stage of the interaction.

Here it is displayed the specificity of behavior of carbcations as amfoteric agents, participating in the reactions with either strong bases – counterions, or weak ones – olefins. Consequently, the reaction of carbcartion with olefin represents unusual composition of electrostatic and orbital interactions, controlled by the presence of counterion.

Making the approach to carbcationic center difficult on the stage of the coordination (the system growth), counterion provides simultaneously the directivity of the further attack, displayed in mutual orientation of components of the system. Their further interaction is accompanied by the energy gain. There exists the basis to suppose, that all the above mentioned holds also for the reaction of olefins with solvately separated ionic pairs, i.e. at the calculation of the medium influence.

The formation of AC in all above considered cases proceeds at direct participation of olefin (ethylene, propylene, isobutylene), oreover playing the role of coordination additive.

Mutual orientation of the complex ($AlCl_3 \cdot H_2O$, $AlEtCl_2 \cdot H_2O$ or $AlCl_4O \cdot H_2O$) and olefin (C_2H_4, C_3H_6, C_4H_8) is accompanied by loosening of π-bond of olefin, activation of O-H bond in H_2O molecule and sufficient increase of Brenstedt acidic strength of $R_nAlCl_{3-n} \cdot H_2O$ complexes (pKa increases from +4.4 to -14÷16).

Proton transfer to olefin is connected with simultaneous formation of counterion ($[AlCl_3 \cdot OH]^{-\delta}$, $[AlCl_4O \Diamond OH]^{-\delta}$, $[AlEtCl_2 \cdot Oh]^{-\delta}$).

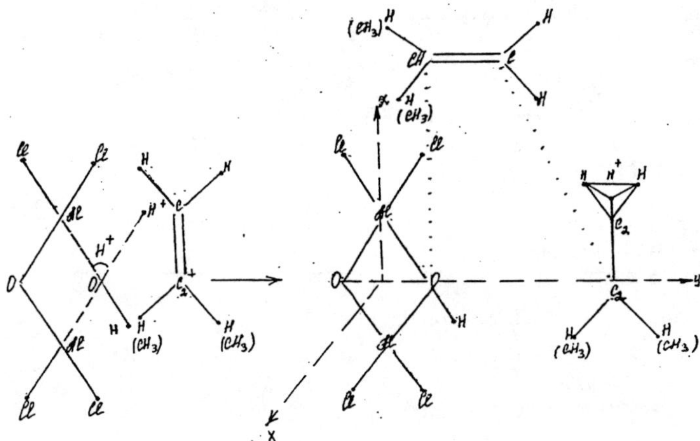

Figure 4.16. The mechanism of initiation and growth at cationic polymerization of ethylene, propylene and isobutylene in presence of aquacomplex of alumoxane ($R_{OC_2^+}$ - reaction coordinate).

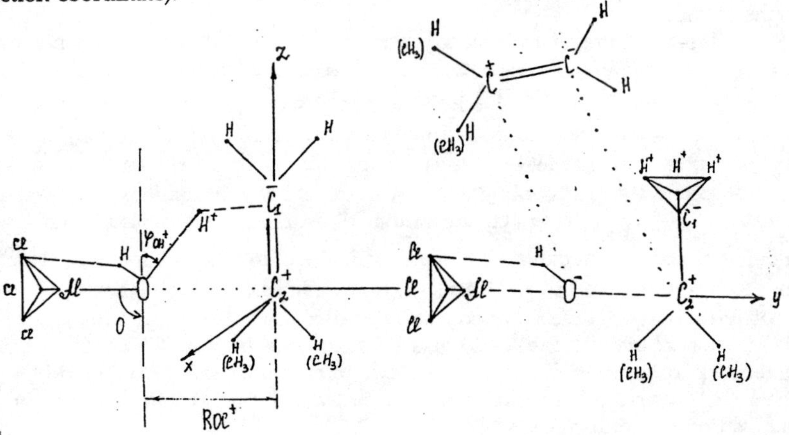

Figure 4.17. The mechanism of initiation and growth at cationic polymerization of ethylene, propylene and isobutylene in presence of $C_2H_5AlCl_2 \cdot H_2O$ ($R_{OC_2^+}$ - reaction coordinates).

Table 4.15. Quantum-chemical characteristics of the interaction of aquacomplexes with olefins.

| No. | Complex | CNDO/2 | | | | | | MINDO/3 |
| | | Initiation | | | | Growth | | Init. |
		φ_{OH^+}	q_{H^+}	$E_g,$ kJ mole	$E_b,$ kJ mole	$E_g,$ kJ mole	$E_b,$ kJ mole	$E_b,$ kJ mole
1.	AlCl3·H2O·C4H8 [11;12]	20	+0.29	+580	+50	+545	+40	+12
2.	AlCl3H2O·C3H6	25	+0.27	+560	+76	+517	+67	+18
3.	AlCl3·H2O·C2H4	40	+0.26	+532	+93	+504	+31	+27
4.	AlCl2HCl·H2O·C4H8	20	+0.30	+615	+52	+560	+37	+12
5.	AlCl2HCl·H2O·C3H6	25	+0.28	+573	+67	+530	+61	+18
6.	AlCl2·HCl·H2O·C2H4	40	+0.26	+542	+82	+518	+76	+28
7.	Al2Cl4O·H2O·C4H8	–	+0.29	+618	+39	+558	+36	+10
8.	Al2Cl4O·H2O·C3H6	–	+0.27	+581	+58	+533	+55	+20
9.	Al2Cl4O·H2O·C2H4	–	+0.26	+551	+80	+515	+73	+26

φ_{OH^+} – angle of olefin attack by proton (H^+);

q_{H^+} – maximal charge on hydrogen atom of water;

E_g – energy gain as a result of the reaction;

E_b – energetic barrier of the reaction of initiation and chain growth.

   Counterions in the composition of AC participate directly in electron transfers, connected with disclosure of the double bond of olefin. Concert (coordinated) mechanism is characteristic for processes of initiation and chain growth at cationic polymerization of olefins.

   Depending on nature of cationic monomers at the excitation of polymerization reaction by the complexes $AlCl_3 \cdot H_2O$, $AlCl_2Et \cdot \cdot H_2O$, $Al_2Cl_4O \cdot H_2O$ it is changing the angle of attack of olefin by initiating particle $H_{+\delta}$ (φ_{OH^+}; Figures 4.16 and 4.17): 20° for isobutylene, 25° for propylene and 40° for ethylene. The charge on H^+ atom of active O-H^+ bond in the molecule of H_2O increases from +0.20 to its maximum value of +0.29; +0.27; +0.26, respectively, according to olefin approaching aquacomplex by the reaction coordinate $RO-_{OC_{(2)}^+}$. It is practically defined by the olefin nature, possessing no dependence on counterion nature. The energy gain, resulting the reaction, (E_g, Table 4.15) depends also on olefin nature and decreases in the sequence C_4H_8, C_3H_6, C_2H_4 on the stage of initiation as well as on the stage of the chain growth at cationic polymerization of olefins. Table 4.15 also presents such important quantum-chemical characteristics of interaction of aquacomplexes of aluminim halogenides with olefins as energetic barriers of the reactions of initiation and chain growth at the polymerization of olefins (E_b).

   Values of E_b (calculated by CNDO/2 method and for the comparison by MINDO/3 method) in the sequence of olefins C_4H_8, C_3H_6, C_2H_4 increase and depend weakly on nature of $[AlCl_3 \cdot OH]^{-\delta}$, $[AlCl_2Et \cdot OH]^{-\delta}$, $[Al_2Cl_4O \cdot OH]^{-\delta}$ counterions.

   The analysis of quantum-chemical parameters (Table 4.15) and the mechanism f interaction of $Al_2Cl_4O \cdot OH$ with olefins showed, that H-initiation is characteristic for alumoxane aquacomplex as for the initial $AlCl_2Et \cdot H_2O$ complex. The calculation coordinates with the experiment, performed before [117], in which weak absorption bends at 2137 and 2183 cm^{-1} (v_{C-D}) were obtained in IR-spectrum of products of oligomerization of diisobutylene C_4H_{16} with the help of $AlCl_2Et \cdot H_2O$ (M_{III}=400). This testifies H-initiation of cationic polymerization of olefins in presence of aquacomplexes of alumoxane.

The structure of AC, formed during the initiation of polymerization of olefins (C_4H_8, C_3H_6, C_2H_4) by complexes $Al_2Cl_4O \cdot H_2O$ and $AlCl_2Et \cdot H_2O$ are presented on the Figures 4.16 and 4.17. Structures of AC1 and AC2 differ from each other by nature of counterion:

$$AC1 - [AlCl_2Et \cdot OH]^{-\delta} ... C^{+\delta}(CH_3)_3$$
$$AC2 - [Al_2Cl_4O \cdot OH]^{-\delta} ... C^{+\delta}(CH_3)_3$$

Despite the differences in the structures of AC, the analysis of parameters of quantum-chemical calculations (Table 4.15), reflecting the energetics of the AC formation (E_g and E_b), points out the possibility of simultaneous existence of active centers of two different types: AC1 and AC2. Theoretical calculations of the formation of AC of two types are proved by the experimental data on the quality of the first rate constants of benzene alkylation $(21.5\pm0.6) \cdot 10^{-4}s^{-1}$, 297°K for the catalyst $AlCl_2Et \cdot H_2O$ and for aquacomplex of alumoxane, previously prepared in benzene $(19.8\pm1.2) \cdot 10^{-4}s^{-2}$, 297°K [117].

Consequently, low value of the acidic strength of aquacomplexes of aluminium halogenides and different forms of their transformations, as well as negative values of ΔE^{H+}, independently on nature of ligand surrounding of Al always suppose the active role of substrate in the performance of acidic-catalytic properties.

4.5. Aquacomplexes of magnesium chlorides.

Combinations of magnesium chlorides with proton-donors, in particular $R_nMgCl_{2-n} \cdot n_1H_2O$ (n=0, 1, 2; n_1=1, 2, 4, 6, 8, 12; R=OH, CH_3, C_2H_5, i-C_3H_7, t-C_4H_9), find their application in industry, medicine, oil-chemical synthesis, olefin polymers production, selective depolymerization and other branches of science and technics [2, 118]. Among these it stands out $MgCl_2 \cdot H_2O$, $MgCl_2 \cdot 2H_2O$ according to their interesting properties and the systems of H_2O-Griniar reactive type. In particular, the distinctive feature of $MgCl_2 \cdot H_2O$ as a catalyst is the ability to polymerize isobutylene up to very high molecular weight, for example. In this case it resigns Al (O~second-C_4H_9)$_3$(BF_3)$TiCl_4$ [2]. The system $MgCl_2 \cdot 6H_2O$ (+19 microelements – bischofite) is characterized by biological activity and by the list of properties stipulating its application in medicine. Compositions of Griniar reactives – water (alcohol) type are interesting from the point of view of Lewis acid inclination to the formation of complexes in dependence on nature of ligand surrounding.

Magnesium chlorides are well-soluble in water (Table 4.16). Hydrates with 1, 2, 4, 6, 8 and 12 molecules of water are extracted from water solutions. The existence of hydrates with 3 and 5 molecules of water is sufficiently disputable.

Quantum-chemical parameters of acidic strength (q_{H+}, pKa) and energy of complex formation (+E_k) of $MgCl_{2-n} \cdot n_1H_2O$ models are presented in the Table 4.17. For more

                    V.A. Babkin, G.E. Zaikov, and K.S. Minsker

comfort in analysis and comparison of parameters the models of aquasystems with magnesium chlorides are subdivided into several groups in dependence on the number of water molecules (n_1) and nature of ligand surrounding of Mg (I-V groups.)

Table 4.16. Solubility of magnesium chlorides in water [118].

Temperature, °C	Solubility of $MgCl_2$, mass %	Solid phase
-10	11.1	Ice
-20	16.0	Ice
-30	19.4	Ice
-33.6	20.6	Ice + $MgCl_2 \cdot 12H_2O$
-20	26.7	$MgCl_2 \cdot 12H_2O$
-16.4	30.6	$MgCl_2 \cdot 12H_2O$
-16.8	31.6	$MgCl_2 \cdot 12H_2O + MgCl_2 \cdot 8H_2O$ (α)
-17.4	32.3	$MgCl_2 \cdot 12H_2O + MgCl_2 \cdot 8H_2O$ (β)
-19.4	33.3	$MgCl_2 \cdot 8H_2O + MgCl_2 \cdot 6H_2O$
-9.6	33.9	$MgCl_2 \cdot 8H_2O(\beta) + MgCl_2 \cdot 6H_2O$
-3.4	34.4	$MgCl_2 \cdot 8H_2O(\alpha) + MgCl_2 \cdot 6H_2O$
0	34.5	$MgCl_2 \cdot 6H_2O$
10	34.9	-"-
20	35.3	-"-
40	36.5	-"-
60	37.9	-"-
80	39.8	-"-
100	42.2	-"-
116.7	46.2	-"-
152.6	49.1	$MgCl_2 \cdot 6H_2O + MgCl_2 \cdot 4H_2O$
181.5	55.8	$MgCl_2 \cdot 4H_2O$
186	56.1	$MgCl_2 \cdot 4H_2O + MgCl_2 \cdot 2H_2O$
		$MgCl_2 \cdot 2H_2O$

Table 4.17. Calculated parameters of acidic strength and complex formation of models of $R_nMgCl_{2-n} \cdot n_1H_2O$ (CNDO/2).

No.	Aquacomplexes	$+E_k$, kJ/mole	q_{H+}	pKa	Δq_{H+}	No. of stage
1.	H_2O	–	+0.14	+15.7	–	
2.	$MgCl_2 \cdot H_2O$	269	+0.20	+7.1	0.15	
3.	$MgCl_2 \cdot 2H_2O$	216	+0.19	+7.7	–	
4.	$MgCl_2 \cdot 4H_2O$	216	+0.17	+10	–	I
5.	$MgCl_2 \cdot 6H_2O$	191	+0.16	+14.4	–	
6.	$MgCl_2 \cdot 8H_2O$	183	+0.16	+14.4	–	
7.	$MgCl_2 \cdot 12H_2O$	167	+0.15	+15.8	–	
8.	$OHMgCl \cdot H_2O$	253	+0.18	+8.5	0.13	
9.	$CH_3MgCl \cdot H_2O$	251	+0.18	+8.5	0.13	
10.	$C_2H_5MgCl \cdot H_2O$	255	+0.18	+8.5	0.13	II
11.	$C_3H_7MgCl \cdot H_2O$	259	+0.18	+8.5	0.13	
12.	$C_4H_9MgCl \cdot H_2O$	253	+0.17	+10	0.12	
13.	$(OH)_2Mg \cdot H_2O$	247	+0.17	+10	0.13	
14.	$(CH_3)_2Mg \cdot H_2O$	244	+0.18	+8.5	0.13	
15.	$(C_2H_5)_2Mg \cdot H_2O$	247	+0.18	+8.5	0.13	III
16.	$(C_3H_7)_2Mg \cdot H_2O$	241	+0.18	+8.5	0.13	
17.	$(C_4H_9)_2Mg \cdot H_2O$	245	+0.18	+8.5	0.13	
18.	$OHMgCl \cdot H_2O$	210	+0.18	+8.5	–	
19.	$CH_3MgCl \cdot H_2O$	215	+0.17	+10	–	
20.	$C_2H_5MgCl \cdot H_2O$	213	+0.17	+10	–	IV
21.	$C_3H_7MgCl \cdot H_2O$	213	+0.18	+8.5	–	
22.	$C_4H_9MgCl \cdot H_2O$	215	+0.17	+10	–	
23.	$(OH)_2Mg \cdot 2H_2O$	203	+0.17	+10	–	
24.	$(CH_3)_2Mg \cdot 2H_2O$	205	+0.18	+8.5	–	
25.	$(C_2H_5)_2Mg \cdot 2H_2O$	203	+0.17	+10	–	V
26.	$(C_3H_7)_2Mg \cdot 2H_2O$	201	+0.17	+10	–	
27.	$(C_4H_9)_2Mg \cdot 2H_2O$	199	+0.18	+8.5	–	

All considered aquacomplexes of Mg halogenides are inclined to the complex formation. The proofs of this statement are high values of energies of complex formation (E_c=167-265 kJ/mole), as the guarantee of the complexes stability. The transition of nonlinked electrons from oxygen atom of water from p_y-orbitals to vacant Sp^2-orbitals of Mg^{2+} takes place at the interaction of water molecule with magnesium chlorides.

For complexes $MgCl_2 \cdot H_2O$ and $RMgCl \cdot H_2O$ it is sufficient the overlapping of d-orbitals of Cl and p_z-orbital of oxygen atom (overlapping integral equals 0.2). The inclination to complex formation of R_nMgCl_{2-n} with H_2O increases in the direction from MgR_2 to $MgCl_2$, i.e. with the increase of Lewis acid strength. This is testified by the character of the change of E_c energy and of part of electron charge, transferred to acceptor (Δq). Independently on nature of the ligand surrounding of Mg all aquacomplexes of magnesium chlorides are characterized by low Brenstedt acidity (pKa=+7.3÷+15.1) (Table 4.17). The highest acidic strength is possessed by complexes $MgCl_2 \cdot H_2O$ (pKa≈+7.3) and $MgCl_2 \cdot 2H_2O$ (pKa≈+7.7). Apparently, this is not a chance, because it is precisely these complexes that are the most efficient catalysts of polymerization [2]. However, electrophilic properties of these complexes cannot be explaimed by the acidic strength only. The additional cause, which explains acidic-catalytic properties of aquacomplexes of magnesium chlorides, is the singularity of coordination of water molecule in relation to $MgCl_2$ (Figure 4.18a). All atoms of the model are disposed in XOY plane (Spd^2-hybridization of the structure).

In the case of $MgR \cdot H_2O$ complexes coordination of water in the models is usual, i.e. the best overlapping of orbitals is observed at the maximum removal of R radicals from each other and from the oxygen atom, which correlates with the calculation, presented in [55] ($R=CH_3$). Apparently, the singularity of water coordination in the models $MgCl_2 \cdot H_2O$, $RMgCl \cdot H_2O$ and $MgCl_2 \cdot 2H_2O$, $RMgCl \cdot 2H_2O$ is explained by additional interaction between d-orbitals of Cl and p_z-orbitals of oxygen atom of water. It is characteristic only for double-valent heteroatom (Mg) of magnesium halogenides. Such coordination of water in the complex increases its sizes up to 0.7÷0.8 nm comparing with the sizes of the corresponding aquacomplexes $Mg(CH_3)_2 \cdot H_2O$ - 0.3÷0.4 nm [55]. These results coincide with the supposition made by J. Kennedy about the fact, that polymerization of isobutylene proceeds under the influence of $MgCl_2$ only in presence of H_2O and is connected with the complex nature of the system, polyisobutylene of extremely high molecular mass being formed [2].

Fuether on (Table 4.17) the models of magnesium chloride with 4, 6, 8 and 12 water molecules are disposed in the group I. In this sequence the energy of complex formation E_c decreases, charges on atoms of hydrogen q_{H+} decrease from +0.20 to +0.15, values of the universal acidity index (pKa) increase from +7.3 to +15.1, that is evident in the general case. Nevertheless, these results prove the validity of calculation schemes and the correctness of comparison of quantum-chemical parameters of aquacomplexes with magnesium halogenides.

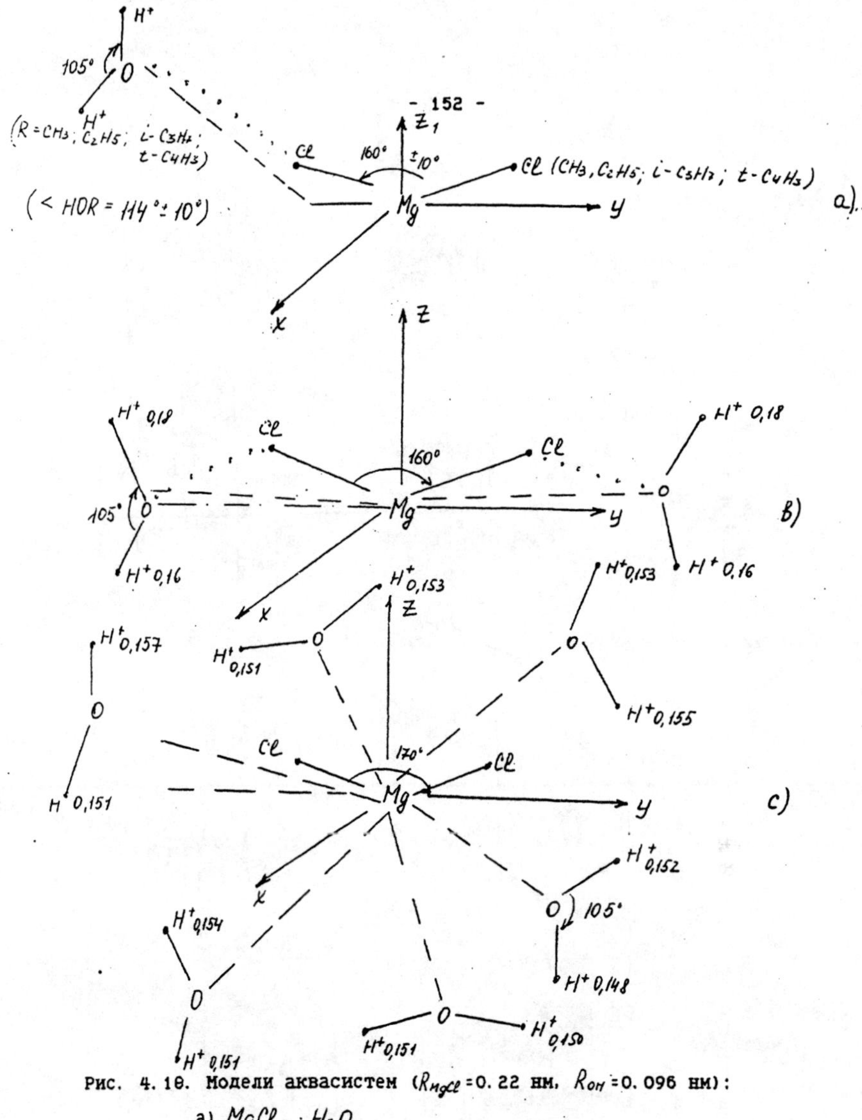

Рис. 4.18. Модели аквасистем (R_{MgCl}=0.22 нм, R_{OH}=0.096 нм):

а) $MgCl_2 \cdot H_2O$

б) $MgCl_2 \cdot 2H_2O$

в) $MgCl_2 \cdot 6H_2O$

Figure 4.18. Models of aquasystems (R_{MgCl}=0.22 nm, R_{OH}=0.096 nm): a) $MgCl_2 \cdot H_2O$; b) $MgCl_2 \cdot 2H_2O$; c)$MgCl_2 \cdot 6H_2O$.

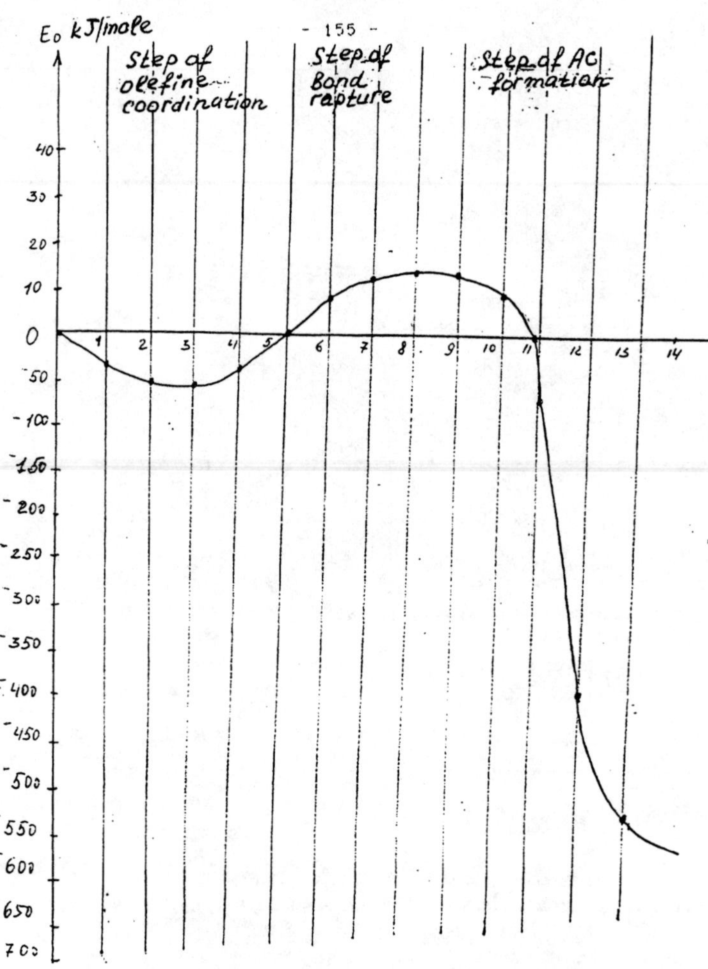

Figure 4.19. Energetic profile of the reaction of $MgCl_2 \cdot H_2O$ interaction with isobutylene (Eint=61 kJ/mole) (CNDO/2).

Figure 4.18a presents the model of $MgCl_2 \cdot 6H_2O$ structure. All six molecules of water of this system are disposed in XOY plane in the same way as in the models with 1, 2 and 4 water molecules. The structure of $MgCl_2 \cdot 6H_2O$ possesses Spd^2-hybridization. As judged with the value of pKa=+14.4 the acidic strength is low. The sixth water molecule completes the electron shell of Mg in XOY plane. Coordination of the seventh molecule in this plane leads to the formation (in local minimums) of different structures of water dimers (linear, bifurcatial or cyclic). The most energetically profitable model in this case is the one possessing Spd^3-hybridization, in which the seventh molecule of water comes out from XOY plane. The rest of H_2O molecules (8, 9, 10, 11 and 12) begin to fill coordinational electron sphere.

The second and the third groups (Table 4.17) consist with $MgR_2 \cdot H_2O$ and $RMgCl \cdot H_2O$ models, which possess Cl ligands substituted by the corresponding radicals - OH, CH_3, C_2H_5, C_3H_7, C_4H_9. Structures of $MgR_2 \cdot H_2O$ ($R=CH_3$) models are close to the ones, presented in [55], after optimization of lengths of the bonds between atoms and of valent angles by the method of variable metrics. It is seen from the Table 4.17 that $MgR_2 \cdot H_2O$ and $RMgCl_2 \cdot H_2O$ as well as $MgCl_2 \cdot H_2O$ are inclined to the complex formation. The energy of complex formation (E_c) of aquacomplexes of magnesium halogenides, the transfer of electron charge to acceptor (Δq), the charge on hydrogen atom of water (q_{H^+}), and consequently the acidic strength (pKa) does not practically depend on nature of the ligand surrounding of Mg. The similar result was obtained for $MgR_2 \cdot 2H_2O$ and $RMgCl \cdot 2H_2O$ complexes also.

The values of pKa of R_nMgCl_{3-n} aquacomplexes are small (+7.1 ÷ +15.1). That is why their activity in electrophilic processes naturally supposes the participation of the reacting bases and coordinated mechanisms of initiation and the chain growth, similarly to aquacomplexes of aluminium chlorides [112]. This was shown on the example of the model of interaction of $MgCl_2 \cdot H_2O$ complex with isobutylene. The distinctive feature of such interaction is in the form of energetic barrier (see the Figure 4.19), overcoming of which by proton (in comparison with $AlCl_3 \cdot H_2O$ complex at isobutylene initiation) is difficult in all probability in consequence of steric obstacles and the influence of big sizes of the forming $[MgCl_2 \cdot OH]$ counterion (because of singularity of H_2O coordination). Apparently, this is the very cause of low rate of polymerization at the initiation $MgCl_2 \cdot H_2O$ [2].

The data obtained at the study of electrophilic depolymerization of polyolefins – polymers of isobutylene (PIB) and butyl caoutchouc (BC) are the experimental prove of quantum-chemical calculations [119]. Degradation of polymers was performed with evaporation of liquid products of depolymerization (for the estimation of conversion) as well as with their returning to the reaction zone (in order to preserve the balance of double bonds in polymeric products). The output of isobutylene in the gas phase is 95-99 reverse % at every conversion of polymeric products (Table 4.18). The rate of elimination of isobutylene at depolymerization of PIB and BC changes in narrow ranges with the decrease of polymeric chain length $\overline{P}_n^o$, but at the same time it depends on the content of end C=C bonds. Total output of the reaction products of the PIB degradation, hydrogenated by 90% (M_n^o=1100, T=583 K, $[MgCl_2 \cdot 2H_2O]$=2 mole/kg) decreases sufficiently (by 7 times) also. As a consequence of predominance of the degradation reaction of polymeric molecules according to chance law, the length of kinetic chain Z of selective depolymerization of PIB and BC is sufficiently shorter that the length of the material chain $\overline{P}_n^o$. It should be mentioned that at BC degradation the values of Z (the length of kinetic chain) approach to the number of monomeric units in block between isopropylic chains. In the case of PIB values of Z are comparable with these parameters for purely thermal process of the polymer degradation (Table 4.18).

Table 4.18. Conversion, isobutylene yield per hour and some kinetic parameters of thermocatalytic degradation of PIB and BC in presence of 1 mole/kg of $MgCl_2 \cdot n_1 H_2O$.

Catalyst	T, K	Conversion, weight %		Yield of IB volumeric %		$Z \pm 10\%$		$v \cdot 10^7$ mole/s $(\pm\%)$		$v_n \cdot 10^7$ mole/s $(\pm 12\%)$	
		PIB	BC	PIB	BC	PIB	BC	PIB	BC	PIB	BC
–	583	1.1	15.4	–	95.5	4.0	6.0	0.8	2.4	0.2	0.4
–	603	8.5	25.0	78.3	96.8	2.5	5.5	3.6	9.0	1.4	1.6
–	623	36.2	48.2	84.5	97.0	2.5	3.8	10.0	15.3	4.0	4.0
$MgCl_2 \cdot n_1 H_2O$	583	12.4	27.4	96.2	96.0	9.9	88.0	6.5	14.0	0.7	0.2
$(n_1=1,2)$	603	81.9	92.0	99.5	96.4	8.1	68.0	13.4	25.0	1.7	0.4
	623	100.0	100.0	99.5	97	7.1	48.0	29.8	35.0	4.2	0.8
$MgCl_2 \cdot 4H_2O$	583	11.4	19.7	90.1	93.1	9.4	72	6.5	7.7	0.7	0.1
	603	36.2	31.6	92.4	92.0	6.8	70	12.4	14.5	1.8	0.2
	623	62.0	35.9	98.6	97	–	56	–	20.4	–	0.4

The results obtained allow to present the scheme of the process of PIB degradation in presence of $MgCl_2 \cdot H_2O$:

1. Initiation by chance law:

$$\sim CH_2 - \underset{\underset{CH_3}{|}}{\overset{\overset{CH_3}{|}}{C}} - CH_2 - \underset{\underset{CH_3}{|}}{\overset{\overset{CH_3}{|}}{C}} \sim \; \rightarrow \; \sim CH_2 + \underset{\underset{CH_3}{|}}{\overset{\overset{CH_3}{|}}{C}} + CH_2 - \underset{\underset{CH_3}{|}}{\overset{\overset{CH_3}{|}}{C}} \sim$$

Disproportioning:

$$\sim CH_2 - \underset{\underset{CH_3}{|}}{\overset{\overset{CH_3}{|}}{C}} - CH_2 - \underset{\underset{CH_3}{|}}{\overset{\overset{CH_3}{|}}{C}} \sim \; \rightarrow \; \sim CH_2 + \underset{\underset{CH_3}{|}}{\overset{\overset{OH_3}{|}}{C}} + CH_2 - \underset{\underset{CH_3}{|}}{\overset{\overset{CH_3}{|}}{C}} \sim$$

2. Initiation according to the end groups law:
a) Formation of catalytically active compound:

$$(MgCl_2)_{ss} \cdot H_2O \rightarrow (MgCl_2)_{22} \rightarrow O \overset{\nearrow H^{+\delta}}{\underset{\searrow H^{+\delta}}{}}$$

$$\sim CH_2 - \underset{\underset{CH_3}{|}}{\overset{\overset{CH_3}{|}}{C}} - CH = \underset{\underset{CH_3}{|}}{\overset{\overset{CH_3}{|}}{C}} + (MgCl_2)_n \cdot OH_s^- H^{+\delta} \rightarrow$$

b)

$$\rightarrow CH_2 - \underset{\underset{CH_3}{|}}{\overset{\overset{CH_3}{|}}{C}} - CH_2 - \underset{\underset{CH_3}{|}}{\overset{\overset{CH_3}{|}}{C}}{}^{+\delta}[(MgCl_2)_n \cdot OH_{solid}]^{-\delta} \rightarrow depolymerization$$

Consequently, experimental data and quantum-chemical calculations show, that the activity of $MgCl_2$ aquacomplexes in thermal depolymerization of PIB and BC is stipulated by the presence of chemically linked water in the catalyst, the charges on hydrogen atoms of which are rigidlylinked with their conversion. As the acidic strength of $R_nMgCl_{2-n}\cdot H_2O$ (pKa$\approx$+7.1$\div$+15.1) is low, the catalytic activity of the complexes of magnesium halogenides is displayed only in presence of the base – acceptor of protons. The presence of the second water molecule in $MgCl_2\cdot 2H_2O$ aquacomplexes does not influence the rate of depolymerization, because temperature of transition of dehydrate into monohydrate (515 K) is sufficiently lower than the reaction temperature (523-623 K). Positive charges on hydrogen atoms in $MgCl_2$ aquacomplexes are in relation with the values of PIB and BC conversion. The decrease of q_{H+} leads to abrupt decrease of the conversion (Table 4.19).

Table 4.19. Some quantum-chemical characteristics of catalysts and the values of conversion of depolymerization of PIB and BC in their presence (T=603 K, [Catalyst]=2 mole/kg, time - 60 min).

No.	Catalyst	E_c^*, kJ/mole	q_{H+}	Conversion, mass %	
				PIB	BC
1.	HCl	–	0.09	8.3	20.1
2.	OHMgCl	–	–	8.5	23.4
3.	$MgCl_2\cdot H_2O$	-265	0.194	100	100
4.		-260	0.185	100	100
5.	$MgCl_2\cdot 2H_2O$	-216	0.165	59.5	45.2
	$MgCl_2\cdot 4H_2O$				

E_c^* - energy of complex formation of chlorides of metal with water per 1 molecule of H2O.

4.6. Aquacomplexes of boron fluorides.

Electron structure and geometry of $R_nBF_{3-n}\cdot H_2O$ complexes are shown on the Figure 4.20. It is interesting the coordination of water according to Lewis R_nBF_{3-n} acid. Additional interaction of one of hydrogen atoms of water with ligand - fluorine - proceeds by means of overlapping of orbitals of these atoms (R_{HF}=0.15 nm). Such interaction is characteristic for the following complexes: $F_3B\cdot H_2O$, $CH_3BF_2\cdot H_2O$, $(CH_3)_2BF\cdot H_2O$, $EtBF_2\cdot H_2O$, $(Et)_2BF\cdot H_2O$. In this case structures of complex acids $R_3B\cdot H_2O$ are close to tetrahedral ones. As judged from the values of charges on hydrogen atoms of water - q_{H+}, and universal index of acidity - pKa (Table 4.20), all the complexes R_nBF_{3-n} possess low acidic strength independent on ligand surrounding of boron. But it is a little bit higher than the one of the previously considdered aquacomplexes, for example $R_nAlCl_{3-n}\cdot H_2O$ and $R_nMgCl_{3-n}\lozenge H_2O$ (Tables 4.17 and 4.20).

The detachment of proton from $R_nBF_{3-n} \cdot H_2O$ is energetically nonprofitable ($\Delta E^{H+} < 0$). That is why it should be expected, that the display of H-activity of $R_nBF_{3-n} \cdot H_2O$ aquacomplexes is connected with the direct interaction with a substrate, and the mechanisms of initiation and chain growth at cationic polymerization of olefins are similar to the ones, which are realized in presence of $R_nAlCl_{3-n} \cdot H_2O$ (see the part 4.4). In particular, this was shown on the example of $BF_3 \cdot H_2O$ - isobutylene system.

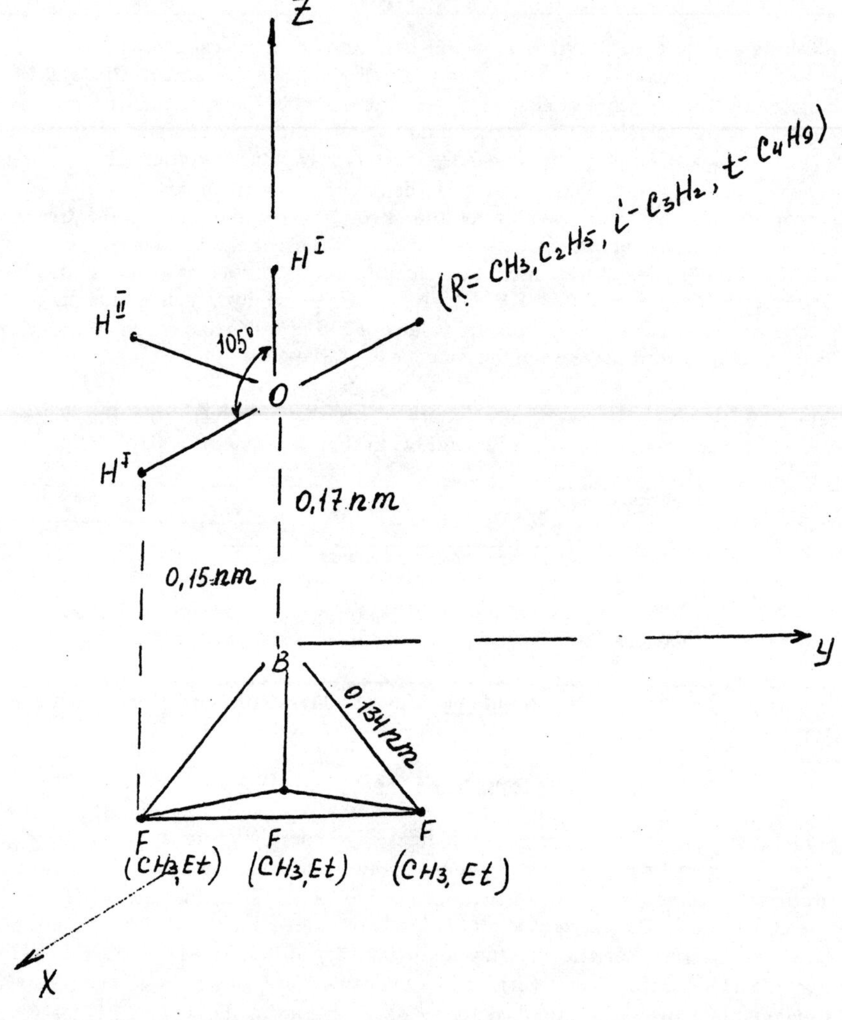

Figure 4.20. Electron structure and geometry of the models of aquacomplexes of boron fluoride (R_{BC}=0.156 - 0.164 nm, <OBF=75°-109°, ΔE^{H+}>420 kJ/mole±60 kJ/mole, <HOR=114°±10°).

Table 4.20. Parameters of acidic strength of aquacomplexes R_nBF_{3-n} (CNDO/2).

No.	Complexes	q_{H^+}	pKa	No. of group
1.	$BF_3 \cdot H_2O$	+0.23	+1.5	
2.	$CH_3BF_2 \cdot H_2O$	+0.22	+4.4	
3.	$(CH_3)_2BF \cdot H_2O$	+0.22	+4.4	I
4.	$(CH_3)_3B \cdot H_2O$	+0.21	+7.1	
5.	$OB_2F_4 \cdot H_2O$	+0.24	+0.1	
6.	$EtBF_2 \lozenge H_2O$	+0.23	+1.5	
7.	$(Et)_2BF \lozenge H_2O$	+0.22	+4.4	II
8.	$(Et)_3B \lozenge H_2O$	+0.22	+4.4	

(For $R=i-C_3H_7$, $t-C_4H_9$ values of $q_{H^+}=+0.21 \div +0.23$, pKa$=+1.0 \div +7.1$, respectively.)

4.7. Complexes of alcohols with aluminium chlorides.

The dependence of electrophilic activity of aluminium chlorides - alcohols complexes on the acidic strength and the ability of alcohol to display proton-donor and (or) carbcationic activity, which is connected with the previous one, [19] allows to determine the interconnection between acidic and catalytic properties of complexes basing on quantum-chemical calculations. Concerning the conditins of the formation and the case for the existence of 1:1 donor-acceptor complexes of alcohol with aluminium chlorides, it is sufficiently proved in [26]. It was obtained that if different alcohols ($CH_3OH \cdot EtOH$, $t-C_4H_8OH$, $C_6H_5CH_2OH$) break Al-C bond in $C_2H_5AlCl_2$, at lower temperature and gradual introduction of alcohol into the reaction zone, for example through diluted solutions of $MnCl_2 \cdot 3CH_3OH$ (0.1 mole/l) in heptane or in water, alcolysis does not practically proceed then. IR-spectrum of heptane solutions of equimolar mixture of $EtAlCl_2$ and EtOH, extracting no ethan at temperatures below 10°C, possesses the displacement of absorption bend of O-H group to the side of lower frequencies with its simultaneous broadening: v_{O-H} ($EtAlCl_2 \cdot EtOH$)=2500-3500 cm instead of v_{O-H} (EtOH)=2246 cm. In analogue with $EtAlCl_2 \cdot H_2O$ complex this fact plays the role of indication of the formation of donor-acceptor complex $EtAlCl_2 \cdot EtOH$(1:1) with polarized O-H bond.

Table 4.21 shows the results of the calculation of individual alcohols. It is seen from this Table, that the orders of $O-H_{(1)}$ bond and the charges on hydrogen atom $q_{H(1)}$ are practically similar for all alcohols. This is coordinated with the experimental values of pKa [120]. The order of $O-C_{(1)}$ bond shows the tendency to decrease at the increase of R' alcohol radical branching. It obtains the minimum value for tret-C_4H_9OH. Positive charge on carbonic atom $C_{(1)}$ and negative charge on oxygen atom increases, respectively. The calculated values of energies of the upper occupied molecular orbital (E_{uoMO}) of alcohols correlates with the values of potentials of ionization (PI) and basicity, known for them [121].

Table 4.22. Quantum-chemical characteristics of protonated alcohols (CNDO/2).

No.	Protonated alcohol	E_o, kJ/mole	P_{OH_1}	P_{OC_1}	q_O	$q_{H(1)}$	$q_{C(1)}$	$\Sigma\Delta q_{H^+}$	$\Delta q_{R'}$	E^*_c, kJ mole	$E^{**}_{H^+}$, kJ mole
1.	$H^+OH_1C_1H_3$	-75980	0.85	0.87	-0.09	+0.36	+0.13	-0.64	-0.29	1090	–
2.	$H^+OH_1C_2H_1C_6H_5$	–	0.86	0.84	-0.13	+0.34	+0.18	–	–	–	–
3.	$H^+OH_1C_1H_2CH_3$	-98810	0.85	0.81	-0.11	+0.34	+0.18	-0.66	-0.32	1120	780
4.	$H^+OH_1C_1H(CH_3)_2$	-121640	0.87	0.74	-0.12	+0.33	+0.21	-0.68	-0.34	1170	810
5.	$H^+OH_1C_1(CH_3)_3$	-144420	0.87	0.71	-0.13	+0.32	+0.23	-0.70	-0.41	1180	830

$*E_c = E_{O(H^+)} + E_{O(R'OH)} - E_{O(H+ROH)}$

$**E_{H^+}$-affinity to proton.

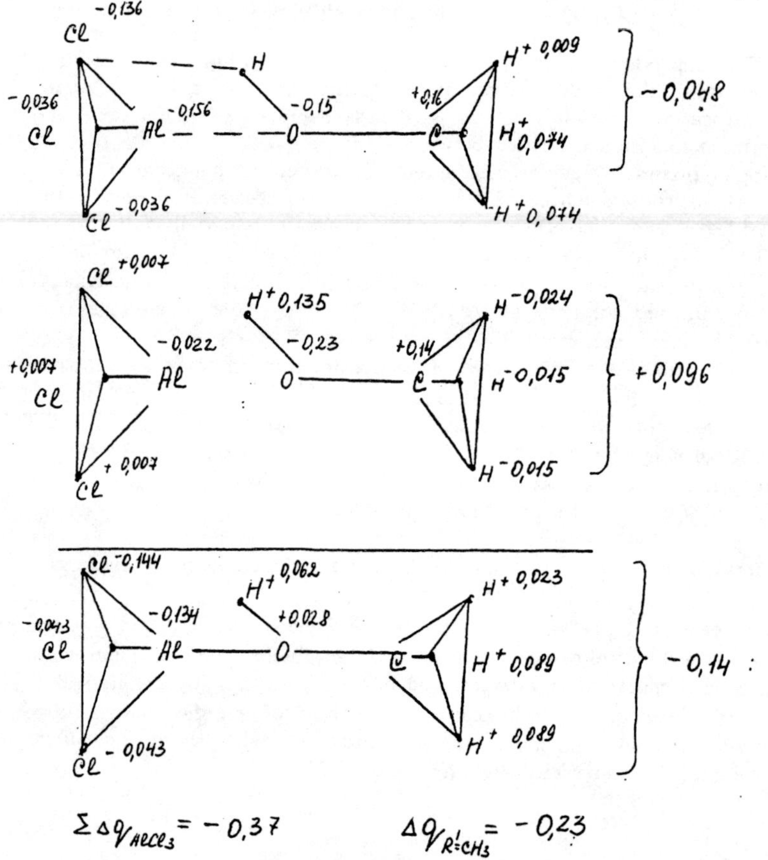

Figure 4.21. Molecular differential diagram according to Holnstein [87] of $AlCl_3 \cdot OHCH_3$ complex (for $AlCl_3 \cdot OHC_2H_5$, $AlCl_3 \cdot OHC_3H_7$, $AlCl_3 \cdot OHC_4H_9$ - similarly) (CNDO/2).

Table 4.23. Calculated data of the complexes of alcohols with carbonium ions (CNDO/2).

No. Alcohol + carbonium ion	$P_{OH(1)}$	$P_{OC(1)}$	$P_{OC(2)}$	q_o	$q_{H(1)}$	$q_{C(1)}$	$q_{C_{(2)}^+}$
1. $H_3C_{(2)}^+ \cdot OH_{(1)}C_{(1)}H_3$	0.85	0.87	0.88	-0.04	+0.34	+0.13	+0.13
2. $H_3C_{(2)}^+ \cdot OH_{(1)}C_{(1)}H(CH_3)_2$	0.88	0.78	0.89	-0.06	+0.32	+0.20	+0.14
3. $H_3C_{(2)}^+ \cdot OH_{(1)}C_{(1)}(CH_3)_3$	0.87	0.70	0.89	-0.09	+0.32	+0.26	+0.13
4. $(CH_3)_3C_{(2)}^+ \cdot OH_{(1)}C_{(1)}H_3$	0.87	0.89	0.70	-0.09	+0.30	+0.13	+0.26
5. $(CH_3)_3C_{(2)}^+ \cdot OH_{(1)}C_{(1)}H(CH_3)_2$	0.88	0.78	0.70	-0.13	+0.27	+0.23	+0.27
6. $(CH_3)_3C_{(2)}^+ \cdot OH_{(1)}C_{(1)}(CH_3)_3$	0.88	0.70	0.70	-0.14	+0.27	+0.27	+0.27

Table 4.24. Quantum-chemical characteristics of the complexes of alcohols with $AlCl_3$ (CNDO/2).

No.	$AlCl_3 \cdot ROH^*$ complex	$E_o,$ kJ/mole	$P_{OH(1)}$	$P_{OC(1)}$	P_{OAl}	q_o	$q_{H(1)}$	$q_{C(1)}$	$\Sigma\Delta q_{AlCl3}$	$\Delta q_{R'}$	δpKa
1.	OH1C1H3	-209860	0.93	0.99	0.31	-0.15	+0.20	+0.16	-0.37	-0.23	7.5
2.	OH1C1H2C6H5	–	0.93	0.94	0.31	-0.17	+0.20	+0.18	–	–	7.5
3.	OH1C1H2CH3	-232660	0.94	0.93	0.31	-0.17	+0.19	+0.19	-0.38	-0.26	9
4.	OH1C1H(CH3)2	-255320	0.94	0.90	0.32	-0.18	+0.18	+0.19	-0.40	-0.29	10.5
5.	OH1C2(CH3)3	-278080	0.94	0.88	0.32	-0.20	+0.17	+0.25	-0.46	-0.40	12

$^*R=CH_3, CH_2C_6H_5, C_2H_5, i\text{-}C_3H_7, t\text{-}C_4H_9.$

Table 4.21. Calculated models of alcohols and some their physico-chemical characteristics (CNDO/2).

No.	Alcohol	$E_o,$ kJ/mole	$P_{OH(1)}$	$P_{OC(1)}$	q_o	$q_{H(1)}$	$q_{C(1)}$	$E_{UOMO},$ kJ mole	PI [116] kJ mole	Basicity, [121]	pKa [120]
1.	$OH_1C_1H_3$	-74890	0.95	1.00	-0.23	+0.14	+0.15	-1460	1050	-2.2	16
2.	$OH_1C_1H_2C_6H_5$	–	0.96	0.99	-0.25	+0.13	+0.17	–	–	–	–
3.	$OH_1C_1H_2CH_3$	-97690	0.96	0.99	-0.24	+0.13	+0.18	-1380	1-1-	–	18
4.	$OH_1C_1H_2CH_3$	-120470	0.96	0.97	-0.26	+0.13	+0.20	-1350	990	-3.2	–
5.	$OH_1C_1H(CH_3)_2$	-143240	0.97	0.94	-0.27	+0.12	+0.20	-1320	970	-4.1	19

Protonated alcohols display no practical change of O-$H_{(1)}$ bond order and the charge on $H_{(1)}$ atom. At the same time the order of O-$C_{(1)}$ bond decreases with the increase of the substitution degree at $C_{(1)}$ carbon atom of alcohol (Table 4.22). Polarization of O-$C_{(1)}$ bond grows in this direction simultaneously. Weakening of O-$C_{(1)}$ bond of alcohols in the sequence from CH_3OH to tret-C_4H_9OH is accompanied by the electrons transfer to H^+ acceptor ($\Sigma\Delta q_{H^+}=-0.64 \div 0.70$), the main contribution to which is made by the transfer of electrons from alcohol radical R($\Delta q_{R'}=-0.29 \div -0.41$). Maximum values $\Sigma\Delta q_{H^+}=-0.70$ and $\Delta q_{R'}=-0.41$ were obtained for the complex of H^+ with tret-C_4H_9OH. These calculated data correlate with the values of E_{uoMO} (Table 4.21) and the affinity to proton E_{H^+} of individual alcohols (Table 4.22). In other words, coordination of alcohols with H^+ allows to search the influence of the nature of alcohol radical on the direction of its fragmentation and, as judged from the values of charges on atoms and the orders of $O-H_{(1)}$ and $O-C_{(1)}$ bonds, to clear up high probability of fragmentation by $O-C_{(1)}$ bond for $tret-C_4H_9OH$.

The data, confirming R^+-activity, were also obtained at the coordination of alcohols with acids weaker than H^+-carbcations C^+H_3 and $C^+(CH_3)_3$ (Table 4.23). As well as in the case of protonated alcohols, the order of $P_{OC_{(1)}}$ bond decreases in the sequence of the complexes being considered, and polarization of $O-C_{(1)}$ bond increases, by maximum degree for $C_{(2)}^+H_3 \cdot OH_{(1)} \cdot C_{(1)}(CH_3)_3$ and $C_{(2)}^+(CH_3)_3 \cdot OH_{(1)}C_{(1)}(CH_3)_3$. Further decrease of the strength of Lewis acid ($AlCl_3$) in the composition of complexes levels the certain measure different in the orders of $O-H_{(1)}$ and $O-C_{(1)}$ bonds for each alcohol and the difference in the order of $O-C_{(1)}$ bond between different alcohols (Table 4.24). Whereas the order of $O-C_{(1)}$ bond decreases by $1\div12\%$ in the complexes with $AlCl_3$ comparing with individual alcohols, $P_{O-C_{(1)}}$ of alcohols in complexes with H^+, C^+H_3, $C^+(CH_3)_3$ decreases by $15\div37\%$. Weakening of $O-C_{(1)}$ bond for the alcohols of individual type in the sequence of acids H^+, $C_{(2)}^+(CH_3)_3$, $AlCl_3$ corresponds to the decrease of orders of coordinational bonds of complexes: 0.87 ($R_{OH_{(1)}}$), 0.70 ($R_{OC_{(2)}^+}$), 0.31 (R_{OAl}). The values of $\Sigma\Delta q_{AlCl_3}$ and Δq_R' (Table 4.24, Figure 4.21) for complexes $AlCl_3$-alcohol are lower than corresponding values of $\Sigma\Delta q_{H^+}$ and Δq_R' for the same alcohols in complexes with H^+ (Table 4.22). However, as for other acids, in the case of application of $AlCl_3$ the highest changes of quantum-chemical parameters in the sequence of studied alcohols are characteristic for $tret-C_4H_9 \cdot OH$.

Table 4.25 presents the results of quantum-chemical calculations of models of alcohols with R_nAlCl_{3-n} (q_{H^+} - charge on hydrogen atom of alcohol, P_{OH} and P_{OC} - the orders of O-H and O-C bonds according to Armstrong [72]). Moreover, this Table presents also the values of pKa - universal acidity index and the activity of alcohols (H^+ and R^+) in complexes, which is predominant in the particular compound. H^+-, R^+-activity should be accepted as follows, that the display of proton-donor ($H^+[R_nAlCl_{3-n} \cdot OR]$) and carbcationic ($R^+[R_nAlCl_{3-n} \cdot OH]$) activities in the corresponding $R_nAlCl_{3-n} \cdot ROH$ complex is equally probable. All the models studied were subdivided into XVII groups. I group – the sequence of individual alcohols: methyl, ethyl, isopropryl and tret-butyl. II - IV groups – complexes of alcohols with acids: H^+, H_3^+C, $(H_3C)_3$, C^+. V-XVII gropus – complexes of alcohols with aluminium halogenides, differing by the substitutors.

Table 4.25 shows, that charges on hydrogen atom of individual alcohols q_{H^+} decrease insufficiently at the increase of the degree of radical branching in the alcohol molecule. This coordinates with the literature values of pKa for indiviual alcohols [120]. The bond order P_{OH} is practically similar for all alcohols. P_{OC_1} possesses the inclination for the decrease of its values and displays, that for methyl and ethyl alcohols it is advisable proton-donor activity (H^+), and for tret-butyl alcohol – carbcationic activity (R^+). In the case of application of isopropyl alcohol as cocatalyst with R_nAlCl_{3-n} H^+ and R^+ activities are displayed with equal probability ($P_{OH}\approx0.96$, $P_{OC}\approx0.97$).

Table 4.25. Quantum-chemical characteristics of alcohols and complexes of alcohols with R_nAlCl_{3-n} (CNDO/2).

No.	Model	q_{H^+}	pKa	P_{OH}	P_{OC}	Activity	No. of group
1.	CH_3OH	+0.14	+16.0	0.96	1.00	H^+	
2.	EtOh	+0.13	+18.0	0.96	0.99	H^+	I
3.	C_3H_7OH	+0.13	+18.0	0.96	0.97	H^+, R^+	
4.	C_4H_9OH	+0.12	+19.0	0.97	0.94	R^+	
5.	H^+CH_3OH	+0.36	-29.0	0.85	0.87	R^+, H^+	
6.	H^+EtOH	+0.34	-24.0	0.85	0.81	R^+	II
7.	$H^+C_3H_7OH$	+0.33	-22.0	0.87	0.74	R^+	
8.	$H^+C_4H_9OH$	+0.32	-20.0	0.87	0.71	R^+	
9.	$H_3C^+ \cdot OHCH_3$	+0.34	-24.0	0.86	0.87	R^+, H^+	
10.	$H_3C^+ \cdot OHEt$	+0.32	-20.0	0.88	0.78	R^+	III
11.	$H_3C^+ \cdot OHC_3H_7$	+0.32	-20.0	0.87	0.73	R^+	
12.	$H_3C^+ \cdot OHC_4H_9$	+0.32	-20.0	0.87	0.70	R^+	
13.	$(CH_3)_3C^+ \cdot OHCH_3$	+0.30	-17.0	0.87	0.89	R^+, H^+	
14.	$(CH_3)_3C^+ \cdot OHEt$	+0.27	-11.0	0.87	0.81	R^+	IV
15.	$(CH_3)_3C^+ \cdot OHC_3H_7$	+0.28	-13.0	0.88	0.78	R^+	
16.	$(CH_3)_3C^+ \cdot OHC_4H_9$	+0.27	-11.0	0.88	0.70	R^+	
17.	$Cl_3Al \cdot OHCH_3$	+0.20	+7.5	0.93	0.99	H^+	
18.	$Cl_3Al \cdot OHEt$	+0.19	+7.7	0.94	0.93	H^+	V
19.	$Cl_3Al \cdot OHC_3H_7$	+0.18	+8.5	0.94	0.90	H^+, R^+	
20.	$Cl_3Al \cdot OHC_4H_9$	+0.17	+12.0	0.94	0.88	R^+	
21.	$Cl_2CH_3Al \cdot OHCH_3$	+0.20	+7.5	0.93	0.99	H^+	
22.	$Cl_2CH_3Al \cdot OHEt$	+0.19	+8.5	0.93	0.95	H^+, R^+	VI
23.	$Cl_2CH_3Al \cdot OHC_3H_7$	+0.16	+7.7	0.94	0.91	R^+	
24.	$Cl_2CH_3Al \cdot OHC_4H_9$	+0.17	+12.0	0.94	0.90	R^+	
25.	$Cl(CH_3)_2Al \cdot OHCH_3$	+0.18	+8.5	0.92	0.98	H^+	
26.	$Cl(CH_3)_2Al \cdot OHEt$	+0.17	+12.0	0.92	0.93	H^+, R^+	VII
27.	$Cl(CH_3)_2Al \cdot OHC_3H_7$	+0.16	+14.4	0.92	0.90	H^+, R^+	
28.	$Cl(CH_3)_2Al \cdot OHC_4H_9$	+0.15	+15.1	0.93	0.89	R^+	
29.	$(CH_3)_3Al \cdot OHCH_3$	+0.17	+12.0	0.95	0.97	H^+, R^+	
30.	$(CH_3)_3Al \cdot OHEt$	+0.16	+14.4	0.94	0.93	H^+, R^+	VIII
31.	$(CH_3)_3Al \cdot OHC_3H_7$	+0.16	+14.4	0.93	0.91	R^+	
32.	$(CH_3)_3Al \cdot OHC_4H_9$	+0.15	+15.1	0.93	0.90	R^+	
33.	$Cl_2EtAl \cdot OHCH_3$	+0.20	+7.5	0.94	0.99	H^+	
34.	$Cl_2EtAl \cdot OHEt$	+0.19	+7.7	0.94	0.94	H^+, R^+	IX
35.	$Cl_2EtAl \cdot OHC_3H_7$	+0.18	+8.5	0.95	0.90	R^+	
36.	$Cl_2EtAl \cdot OHC_4H_9$	+0.18	+8.5	0.94	0.90	R^+	

37.	$Cl(Et)_2Al \cdot OHCH_3$	+0.17	+12.0	0.93	0.98	H^+	
38.	$Cl(Et)_2Al \cdot OHEt$	+0.16	+14.4	0.95	0.94	H^+, R^+	X
39.	$Cl(Et)_2Al \cdot OHC_3H_7$	+0.17	+12.0	0.94	0.91	R^+	
40.	$Cl(Et)_2Al \cdot OHC_4H_9$	+0.15	+15.1	0.92	0.89	R^+	
41.	$(Et)_3Al \cdot OHCH_3$	+0.17	+12.0	0.93	0.97	H^+	
42.	$(Et)_3Al \cdot OHEt$	+0.16	+14.4	0.92	0.93	H^+, R^+	XI
43.	$(Et)_3Al \cdot OHC_3H_7$	+0.16	+14.4	0.93	0.90	R^+	
44.	$(Et)_3Al \cdot OHC_4H_9$	+0.15	+15.1	0.92	0.89	R^+	
45.	$Cl_2C_3H_7Al \cdot OHCH_3$	+0.20	+7.5	0.94	0.99	H^+	
46.	$Cl_2C_3H_7Al \cdot OHEt$	+0.19	+7.7	0.94	0.93	H^+, R^+	XII
47.	$Cl_2C_3H_7Al \cdot OHC_3H_7$	+0.19	+7.7	0.94	0.90	R^+	
48.	$Cl_2C_3H_7Al \cdot OHC_4H_9$	+0.18	+8.5	0.94	0.89	R^+	
49.	$Cl(C_3H_7)_2Al \cdot OHCH_3$	+0.17	+12.0	0.93	0.99	H^+	
50.	$Cl(C_3H_7)_2Al \cdot OHEt$	+0.16	+14.4	0.92	0.93	H^+, R^+	XIII
51.	$Cl(C_3H_7)_2Al \cdot OHC_3H_7$	+0.16	+14.4	0.94	0.91	R^+	
52.	$Cl(C_3H_7)_2Al \cdot OHC_4H_9$	+0.15	+15.1	0.94	0.90	R^+	
53.	$(CH_3)_3Al \cdot OHCH_3$	+0.17	+12.0	0.93	0.98	H^+	
54.	$(CH_3)_3Al \cdot OHEt$	+0.17	+12.0	0.93	0.94	H^+, R^+	XIV
55.	$(CH_3)_3Al \cdot OHC_3H_7$	+0.16	+14.4	0.94	0.94	H^+, R^+	
56.	$(CH_3)_3Al \cdot OHC_4H_9$	+0.15	+15.1	0.93	0.90	R^+	
57.	$Cl_2C_4H_9Al \cdot OHCH_3$	+0.20	+7.5	0.94	0.98	H^+	
58.	$Cl_2C_4H_9Al \cdot OHEt$	+0.20	+7.5	0.94	0.94	H^+, R^+	XV
59.	$Cl_2C_4H_9Al \cdot OHC_3H_7$	+0.18	+8.5	0.94	0.90	R^+	
60.	$Cl_2C_4H_9Al \cdot OHC_4H_9$	+0.17	+12.0	0.93	0.88	R^+	
61.	$Cl(C_4H_9)_2Al \cdot OHCH_3$	+0.17	+12.0	0.93	0.99	H^+	
62.	$Cl(C_4H_9)_2Al \cdot OHEt$	+0.16	+14.4	0.93	0.94	H^+, R^+	XVI
63.	$Cl(C_4H_9)_2Al \cdot OHC_3H_7$	+0.15	+15.1	0.92	0.90	R^+	
64.	$Cl(C_4H_9)_2Al \cdot OHC_4H_9$	+0.15	+15.1	0.93	0.89	R^+	
65.	$(C_4H_9)_3Al \cdot OHCH_3$	+0.17	+12.0	0.93	0.99	H^+	
66.	$(C_4H_9)_3Al \cdot OHEt$	+0.16	+14.4	0.93	0.94	H^+, R^+	XVII
67.	$(C_4H_9)_3Al \cdot OHC_3H_7$	+0.15	+15.1	0.92	0.90	R^+	
68.	$(C_4H_9)_3Al \cdot OHC_4H_9$	+0.15	+15.1	0.93	0.92	H^+, R^+	

In the complexes of alcohols with strong acids H^+, H_3C^+, $(H_3C)_3C^+$ (II-VI groups, Table 4.25) the charge on hydrogen atom in the alcohol molecules q_{H^+} increases sharply up to $+0.27 \div +0.36$. This is corresponded to the sufficient increase of the acidic strngth of the alcohol complexes with H^+, H_3C^+, $(H_3C)_3C^+$ up to pKa=$-11 \div -29$, comparing with individual alcohols (pKa$\approx +16 \div +19$). The bond order P_{OH} does not depend on nature of the alcohol radical and on acidic strength of the complexes of alcohols of the II-VI groups. It equals $P_{OH} \approx 0.86 \div 0.88$, that is smaller than the values $P_{OH} \approx 0.96 \div 0.97$ characteristic for individual alcohols. In this case it is observed more sufficient dependence of P_{OC} on the branching of alcohol radical, than for the I group of alcohols. That is why complexes of the II-IV groups are inclined to preferable display of carbcationic activity (R^+).

Complexes of alcohols with aluminium compounds from V-XIII groups (Table 4.25) are characterized by low acidic strength and exist in the range of pKa from +7.1 to +15.1 ($q_{H^+}\approx+0.21\div+0.15$, respectively). In this case the orders of bonds P_{OH} do not practically change. P_{OC} decreases in the sequence CH_3OH, $EtOH$, C_4H_9OH in dependence on the alcohol nature, but in less degree than in the case of application of complexes of alcohols with H^+, H_3C^+, $(H_3C)_3C^+$. Here the values of P_{OH} and P_{OC} are levelled in a certain measure. Alongside with the acidic strength other factors also influence the direction of the alcohols fragmentation, linked with Lewis acids. In particular, these factors are specific interactions between components, the nature of anion and accepting bases, etc. [122]. In some cases these are precisely the factors that define the direction of alcohol fragmentation. All complexes in the V-XVII groups possess low Brenstedt acidity with no dependence on the nature of ligand surrounding of Al. However, it follows from the comparison of two sequences of pKa of complexes, which are: the ones with a single chlorine atom substituted (V, VI, IX, XII, XV groups), and the models with two or three chlorine atoms substituted by the corresponding radicals – CH_3, Et, C_3H_7, C_4H_9 (VII, VIII, X, XI, XIII, XIV, XVI, XVII – groups) – that the first group of complexes represents stronger acidic compounds. In this connection the probability of initiation of electrophilic polymerization by complexes of $R_2AlCl \cdot ROH$ and $R_3Al \cdot ROH$ types is low, because in some cases pKa=+15.1$\div$+14.4 for single models, for example $AlCl(Et)_2 \cdot OHC_4H_9$, $AlCl(C_3H_7)_2 \cdot OHC_3H_7$, $Al(C_4H_9)_3 \cdot OHEt$, etc., come close to pKa$\approx$+16$\div$+19 for individual alcohols. As it is known, individual alcohols are not the initiators of cationic processes.

As judged from the values of P_{OH} and P_{OC} in $R_nAlCl_{3-n} \cdot ROH$ methyl alcohol is characterized by proton-donor activity (H^+), and carbcationic activity (R^+) is preferable for isopropyl and tret-butyl alcohols. In the case of ethyl alcohol the display of H^+ and R^+ activities is equally probable. Moreover, during the analysis of the results of quantum-chemical calculations of $R_nAlCl_{3-n} \cdot ROH$ models it was observed the following fact. Complexes $R_3A \cdot ROH$, in which radical structure of ligand surrounding coincides with the structure of alcohol radical (i.e. complexes $(CH_3)_3Al \cdot OHCH_3$, $(Et)_3Al \cdot OHEt$, $(C_3H_7)_3Al \cdot OHC_3H_7$, $(C_4H_9)_3Al \cdot OHC_4H_9$), possess characteristic equal probability of the display of proton-donor and carbcationic activities ($P_{OH}\approx P_{OC}$).

Despite the increase of the acidic strength of alcohol as a consequences of R_nAlCl_{3-n} complex formation, the detachment of proton in alcohols is energetically nonprofitable (negative values of ΔE^{H^+}=-390$\div$-590 kJ/mole). Selectivity of action of the complexes of this type occupies the intermediate position in comparison with the complexes $R_nAlCl_{3-n} \cdot H_2O$ and $R_nAlCl_{3-n} \cdot HCl$ (Table 4.26). Evidently, considering acidic-catalytic properties of $R_nAlCl_{3-n} \cdot ROH$, it is necessary to take into account some features, in particular, low stability, the existence of several ways of transformation, including processes with HCl formation, the probability of the display of carbcationic activity by alcohols along with proton-donor one [122].

To put it differently, the behavior of $R_nAlCl_{3-n} \cdot ROH$ complexes is more complicated, than of $R_nAlCl_{3-n} \cdot HCl$ and $R_nAlCl_{3-n} \cdot H_2O$ ones.

Apparently, $ROH \cdot AlR_nCl_{3-n}$ complexes display their activity (H^+ or R^+) in cationic polymerization (in analog to R_nAlCl_{3-n} aqua-complexes) under the direct participation of olefin, which is caused by their low acidic strength.

In this connection it is sufficiently interesting the study of the process of interaction of complexes R_nAlCl_{3-n} alcohol with olefins (ethylene, propylene, isobutylene) as well as the formation of AC and setting the dependence of AC structure on the nature of alcohol and olefin. The dynamics of interaction of $EtAlCl_2 \cdot R'OH$ complex ($R'=CH_3$, t, $i\text{-}C_3H_7$, $t\text{-}C_4H_9$) with isobutylene at the alcohol displaying in complexes proton-donor (H^+) as well as carbcationic activity (R^+) was investigated for the purpose of clearing up the role of chemical structure of alcohol molecule in the process of AC formation. The calculation of two different ways of the reaction was performed for this aim:

$$
\begin{matrix}
\text{Cl} \quad \text{H} \\
\text{C}_2\text{H}_5\text{Al}\cdot\text{O} \quad +\text{C}_4\text{H}_8 \\
\text{Cl} \quad \text{CH}
\end{matrix}
\begin{bmatrix}
\text{Et} \\
\text{i}-\text{C}_3\text{H}_7 \\
\text{t}-\text{C}_4\text{H}_9
\end{bmatrix}
\longrightarrow
\begin{array}{l}
\text{I} \quad \text{H}(\text{C}_4\text{H}_8)^+[\text{EtAlCl}_2\cdot\text{OCH}_3]^- \begin{bmatrix}\text{Et, i}-\text{C}_3\text{H}_7 \\ \text{t}-\text{C}_4\text{H}_9\end{bmatrix} \\[2em]
\text{II} \quad \text{CH}_3(\text{C}_4\text{H}_8)^+[\text{EtAlCl}_2\cdot\text{OH}]^- \begin{bmatrix}\text{Et} \\ \text{i}-\text{C}_3\text{H}_7 \\ \text{t}-\text{C}_4\text{H}_9\end{bmatrix}
\end{array}
$$

Table 4.26. Oligomerization of mixture of C_4 hydrocarbons in presence of complexes of aluminium chlorides-alcohols type [109].

No.	Catalyst	Composition of mixture of C_3-C_4 hydrocarbons before the reaction, mass %				Conversion of $i\text{-}C_4H_8\%$	Selectivity of $i\text{-}C_4H_8,\%$
		C_3H_6	H-iso C_4H_{10}	iso C_4H_8	α, β C_4H_8		
1.	$EtAlCl_2\cdot OHC_6H_5$	3.0	55.5	37.9	3.5	95	60
2.	$EtAlCl_2\cdot OHC_6H_5$	0.6	55.1	14.9	29.5	90	79
3.	$C_2H_5AlCl_2\cdot OHCH_3$	0.6	55.1	14.9	29.5	95	74

Figure 4.22 presents the mechanism of interaction of $EtAlCl_2\cdot R'OH$ (CH_3OH, EtOH, C_3H_7OH, C_4H_9OH) complexes with isobutylene at the complex displaying H^+-activity (I way of the reaction). Optimization of geometry of the complexes was performed by the angle of olefin attack by proton (φ_{OH+}), the angle characterizing orientation of $EtAlCl_2\cdot OR'$ fragments according to isobutylene (α), the angle of the forming tetrahedron of carbcation (γ), the length of olefin bond ($R_{\overline{C}-C\cdot}$), the length of $O\text{-}H^+$ bond (R_{OH+}), displacement of $EtAlCl_2\cdot R'OH$ molecule according to Z axis (R_z). The distance R_{OC+} was selected as the reaction coordinate. The most important quantum-chemical and geometric characteristics of the $EtAlCl_2\cdot R'OH$ interaction with C_4H_8 are presented in the Table 4.27 (the angle of isobutylene attack by proton H^+ (φ_{OH+}), q_{H+} – the charge on H^+, heat energy of the reaction E_{int} and activation energy (E_a)). In the case of display of H^+-activity by catalytic complexes the angle of attack of isobutylene by initiating particles φ_{OH+} equals 15°. The characteristic features of the stage of the coordination of olefin (C_4H_8) with $EtAlCl_2\cdot ROH$ are the deflection of $[EtAlCl_2\cdot ROH]^-$ fragment of the complex by the angle $\alpha=80°\div90°$ and the orientation of

O_1-H^+ bond in space to the side of isobutylene. The charge on H^+ atom in the alcohol molecule increases up to it maximum value +0.32, +0.30, +0.29, +0.28, respectively, in dependence on chemical structure of the alcohol (see Table 4.27) as the complex $EtAlCl_2 \cdot ROH$ is approached by isobutylene by the reaction coordinate R_{OC}. In this connection the values of the universal index of acidity pKa are -20÷-13 according to the correlation dependence $pKa=f(q_{H^+})$ [4]. Heat effect of the reaction E_{int} in the sequence of the studied alcohols in the complexes is inclined to increase. The activation energy E_a in the same sequence also increases, which is confirmed by both CNDO/2 and MINDO/3 methods (Table 4.27).

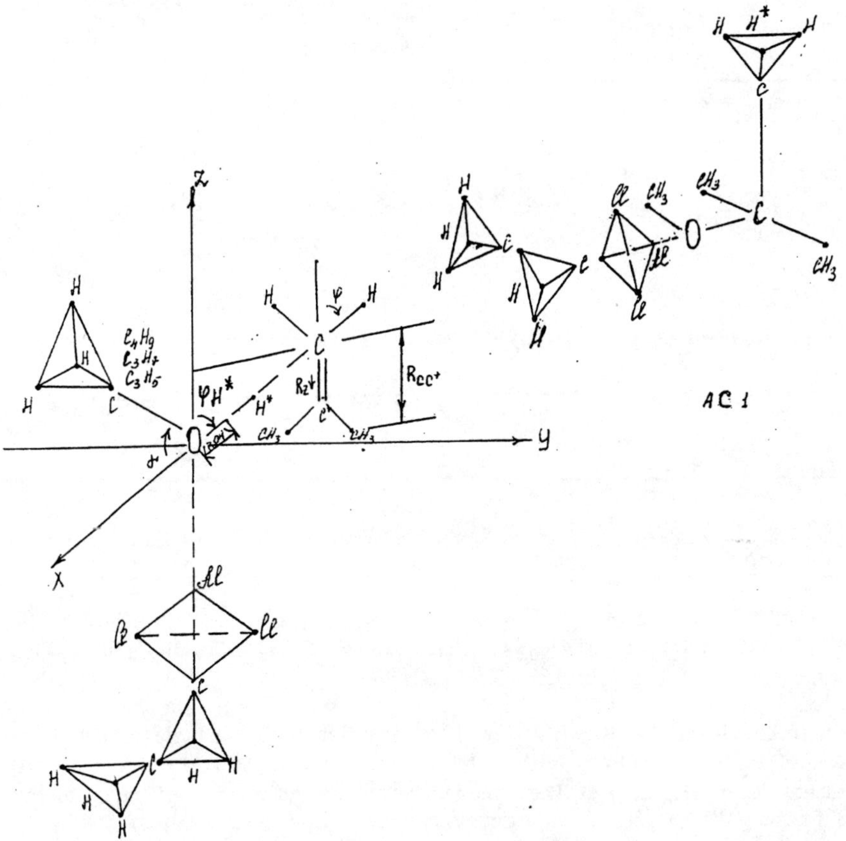

Figure 4.22. Interaction of the complexes of $C_2H_5AlCl_2 \cdot OHCH_3$, $C_2H_5AlCl_2 \cdot C_2H_5OH$, $C_2H_5AlCl_2 \cdot OHC_3H_7$ and $C_2H_5AlCl_2 \cdot OHC_4H_9$ with isobutylene at performing proton-donor activity by alcohols.

102 V.A. Babkin, G.E. Zaikov, and K.S. Minsker

Table 4.27. Quantum-chemical characteristics of interaction of complexes $C_2H_5AlCl_2 \cdot ROH$ with isobutylene.

No.	Model	Activity	CNDO/2						MINDO/3	
			$\varphi_{01}H^+$ grad	$\varphi_{01}C_2$ grad	$q_{H_9^+}$	$q_{C_2^+}$	E_{int} kJ mole	E_a kJ mole	E_a kJ mole	
1.	$C_2H_5AlCl_2 \cdot CH_3OH \cdot C_4H_8$	H^+	15	–	+0.32	–	+591	+42	+10	
2.	$C_2H_5AlCl_2 \cdot C_2H_5OH \cdot C_4H_8$	H^+	15	–	+0.30	–	+597	+51	+13	AC1
3.	$C_2H_5AlCl_2 \cdot C_3H_7OH \cdot C_4H_8$	H^+	15	–	+0.29	–	+604	+67	+19	
4.	$C_2H_5AlCl_2 \cdot C_4H_9OH \cdot C_4H_8$	H^+	15	–	+0.28	–	+609	+75	+21	
1.	$C_2H_5AlCl_2 \cdot CH_3OH \cdot C_4H_8$	R^+	–	20	–	+0.26	+611	+79	+24	
2.	$C_2H_5AlCl_2 \cdot C_2H_5OH \cdot C_4H_8$	R^+	–	20	–	+0.27	+613	+63	+15	AC2
3.	$C_2H_5AlCl_2 \cdot C_3H_7OH \cdot C_4H_8$	R^+	–	10	–	+0.28	+617	+49	+12	
4.	$C_2H_5AlCl_2 \cdot C_4H_9OH \cdot C_4H_8$	R^+	–	10	–	+0.28	+623	+44	+9	

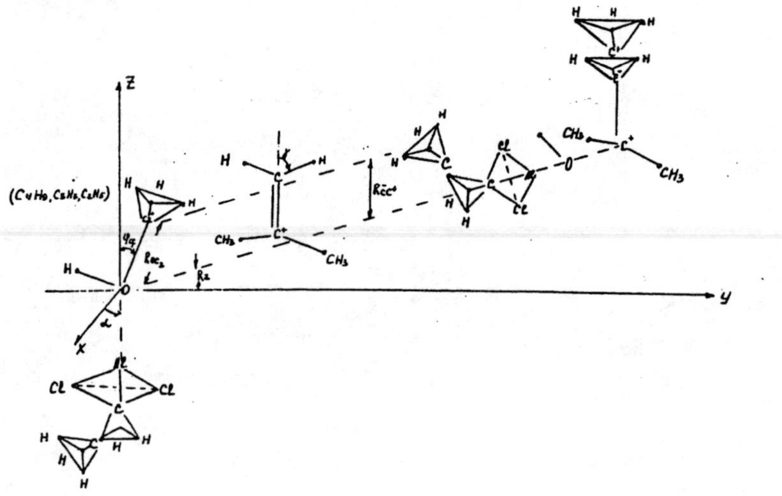

Figure 4.23. Interaction of the complexes of $C_2H_5AlCl_2 \cdot OHCH_3$, $C_2H_5AlCl_2 \cdot OHC_2H_5$, $C_2H_5AlCl_2 \cdot OHC_3H_7$ and $C_2H_5AlCl_2 \cdot OHC_4H_9$ with isobutyleneat performing carbcationic activity by alcohols.

The mechanism of $EtAlCl_2 \cdot ROH$ complex interction with isobutylene at alcohol displaying carbcationic activity in the complexes with $EtAlC_2$ (II way of the reaction) is presented on the Figure 4.23. The same Figure presents geometric parameters of the model $EtAlCl_2 \cdot ROH \cdot C_4H_8$, according to which the optimization was performed (six-dimensional space of the optimizing parameters), R_{OC^+} – the reaction coordinate. The comparison of the values of energetic heat effects E_{int} at alcohols displaying proton-donor and carbcationic activities in the complexes (Table 4.27) shows that the character of their changes is similar for I and II ways of the reaction proceeding. The values of E_a increase for I direction of the reaction and decrease for II direction in unitypical sequence of alcohols CH_3OH, $EtOH$, C_3H_7, C_4H_9OH in the complexes with $EtAlCl_2$.

The structure of AC at the polymerization of isobutylene on the basis of $EtAlCl_2 \cdot ROH$ complexes at the alcohols displaying proton-donor and carbcationic activities are presented on the Figure 4.22 (AC1) and the Figure 4.23 (AC2). They represent polarized intermediates (C^+-0 bond length is a little bit longer than for covalent one), which may be considered as the border case of the ionic pair.

AC1 is the structure, characteristic for the reaction performing according to I direction in dependence on nature of alcohol in the complexes (H^+-activity):

1. $[C_2H_5AlCl_2 \cdot OCH_3]^{-\delta} \ldots C^{+\delta}(CH_3)_3$

2. $[C_2H_5AlCl_2 \cdot OC_2H_5]^{-\delta} \ldots C^{+\delta}(CH_3)_3$ AC1

3. $[C_2H_5AlCl_2 \cdot OC_3H_7]^{-\delta} \ldots C^{+\delta}(CH_3)_3$

4. $[C_2H_5AlCl_2 \cdot OC_4H_9]^{-\delta} \ldots C^{+\delta}(CH_3)_3.$

AC2 is the structure, characteristic for the reaction performing according to II direction (R^+-activity), respectively:

1. $[C_2H_5AlCl_2 \cdot OH]^{-\delta} \ldots C^{+\delta}(CH_3)_2C_2H_5$

2. $[C_2H_5AlCl_2 \cdot OH]^{-\delta} \ldots C^{+\delta}(CH_3)_2C_3H_7$

3. $[C_2H_5AlCl_2 \cdot OH]^{-\delta} \ldots C^{+\delta}(CH_3)_2C_4H_9$ AC2

4. $[C_2H_5AlCl_2 \cdot OH]^{-\delta} \ldots C^{+\delta}(CH_3)_2C_5H_{11}.$

Compositions of AC1 and AC2 differ sufficiently from each other by the structure of both counterions and carbcations. In the case of AC1 the structure of carbcation does not change in dependence on the nature of alcohol in the complexes, but the structure of counterion changes. In opposite, in the case of AC2 the structure of counterion is unchangeable, and the structure of carbcation changes.

The analysis of the data of quantum-chemical calculations (E_a and E_{int}, Table 4.27), which reflect the energetics of AC formation, points out the probability of simultaneous existence of the active centers of two different types: AC1 and AC2

1. $[C_2H_5AlCl_2 \cdot OCH_3]^{-\delta} \ldots C^{+\delta}(CH_3)$ and $[C_2H_5AlCl_2 \cdot OH]^{-\delta} \ldots C^{+\delta}(CH_3)_2C_2H_5$

2. $[C_2H_5AlCl_2 \cdot OC_2H_5]^{-\delta} \ldots C^{+\delta}(CH_3)_3$ and $[C_2H_5AlCl_2 \cdot OH]^{-\delta} \ldots C^{+\delta}(CH_3)_2C_3H_7$

3. $[C_2H_5AlCl_2 \cdot OC_3H_7]^{-\delta} \ldots C^{+\delta}(CH_3)_3$ and $[C_2H_5AlCl_2 \cdot OH]^{-\delta} \ldots C^{+\delta}(CH_3)_2C_4H_9$

4. $[C_2H_5AlCl_2 \cdot OC_4H_9]^{-\delta} \ldots C^{+\delta}(CH_3)_3$ and $[C_2H_5AlCl_2 \cdot OH]^{-\delta} \ldots C^{+\delta}(CH_3)_2C_5H_{11}.$

The data of IR-spectra of products of C_8 oligomerization in presence of $C_2H_5AlCl_2$-CD_3OH, UV and PMR-spectra of $C_2H_5AlCl_2$-$C_2H_5CH_2OH$, GLC (gel-liquid chromatography) of $C_2H_5AlCl_2$-C_4H_9OH, comparison of values of MM of the products, the degree of their nonsaturation and the functional groups content [26] allows to make a conclusion, that te ratio AC1/AC2 in reactionary systems is defined by stability of carbcations, which depends sufficiently on chemical structure of alcohol. Energetic characteristics of the formation of AC1 and AC2 (Table 4.27) allow to set the dependence: for methyl alcohol AC1/AC2>1, i.e. the amount of active centers of I type is larger than of II one; for ethyl alcohol AC1/AC2=1, that means equal probability of formation of AC of the first and the second type; for isopropyl and tret-butyl alcohols the ratio AC1/AC2<1, that means predominant initiation by R^+ cation, the amount of AC2 type is larger.

Consequently, it is evident that acidic-catalytic properties of the complexes of $C_2H_5AlCl_2 \cdot ROH$ type are displayed similar to these of R_nAlCl_{3-n} aquacomplexes only at the direct participation of the molecule of monomer, isobutylene in the present case.

Thus, alcohols in $C_2H_5AlCl_2 \cdot ROH$ complexes at the initiation of cationic polymerization of isobutylene display proton-donor activity as well as the carbcationic one. In this case it takes place coordinational mechanisms of initiation of polymerization and active centers of two types AC1 and AC2 are formed. It is characteristic, that their ratio AC1/AC2 in the reactionary system depends on chemical structure of the alcohol, participating in the composition of initiating complex.

To calculate the influence of olefin role in the formation of AC it was studied the dynamics of the interaction of $C_2H_5AlCl_2 \cdot C_2H_5OH$ complex with olefins – C_2H_4, C_3H_6 and C_4H_8 at the display by ethyl alcohol of both proton-donor (H^+) and carbcationic (R^+) activity in complexes. Calculation of the ways of the reaction was performed for this purpose:

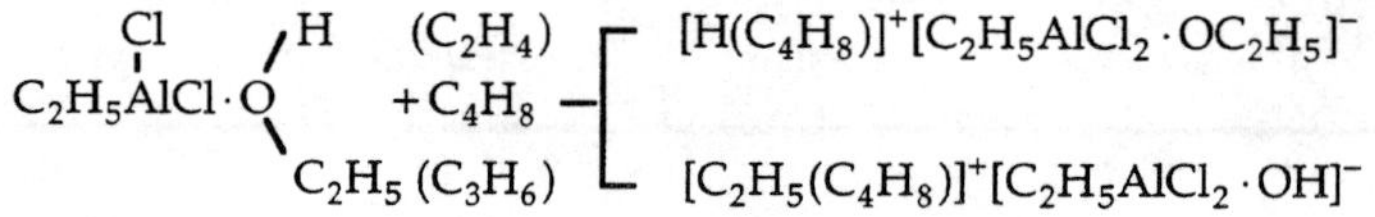

$$\underset{C_2H_5AlCl \cdot O}{\overset{Cl}{|}}\diagup^{H}_{\diagdown C_2H_5}\quad \overset{(C_2H_4)}{\underset{(C_3H_6)}{+C_4H_8}} \quad \left[\begin{array}{c} [H(C_4H_8)]^+[C_2H_5AlCl_2 \cdot OC_2H_5]^- \\[2mm] [C_2H_5(C_4H_8)]^+[C_2H_5AlCl_2 \cdot OH]^- \end{array} \right.$$

Scheme IV

The mechanism of the interaction of $C_2H_5AlCl_2 \cdot C_2H_5OH$ with olefins (ethylene, propylene, isobutylene) at the display of H^+-activity (I way of the reaction) is presented on the Figure 4.24. Optimization of geometry of complexes was performed according to angles φ_{OH9}, α, γ and interatomic distances R_{OH9}, R_z, $R_{C_2C_2O}$ (six-dimensional space of optimizing parameters). The most important quantum-chemical and geometric characteristics of interaction of ethyl alcohol with olefins (the angle of olefin attack by proton (H_9^+) - φ_{OH9}, the deflection angle of forming counterion $[C_2H_5AlCl_2 \cdot OC_2H_5]^-$ in relation with isobutylene α, the charge on H_9^+ hydrogen atom q_{H+}, heat energies of the reaction E_{int} and activation energgies of the reactions E_a) are presented in the Table 4.28. For independent control of the correctness of the calculation scheme the characteristics of the interaction of $C_2H_5AlCl_2 \cdot C_2H_5OH$ complex with C_2F_4 monomer are presented also. This monomer is unable to polymerize according to cationic mechanism. In this case the initiation of the process of polymerization takes no place. The attachment of the initiating H_9^+ particle to C_2F_4 led to the increase of energy of the complex $C_2H_5AlCl_2 \cdot C_2H_5OH \cdot C_2F_4$ and to the energy losses resulting the reaction of 32 kJ/mole (at the display of H^+-activity) and 27 kJ/mole (at the display of R^+-activity) (Table 4.28).

Depending on chemical structure of olefins at the initiation of cationic polymerization by complex catalysts $C_2H_5AlCl_2 \cdot OHC_2H_5$ at the display of proton-donor activity by ethyl alcohol the angle of attack of initiating particle H_9^+- φ_0 changes in the following way: 15° for isobutylene, 25° for propylene and 35° for ethylene (Table 4.28). At the stage of initiation the behavior of all olefins is characterized by deflection of the fragment $[C_2H_5AlCl_2 \cdot OC_2H_5]^-$ of the complex (it does not change on further stages of the reaction). $O_2 - H_9^+$ bond orients to the side of the monomer. The charge on H_9 hydrogen atom of the alcohol increases up to its maximum value +0.30 (+0.29, +0.27, respectively) as isobutylene (propylene, ethylene) approaches to $C_2H_5AlCl_2 \cdot OHC_2H_5$ complex by the reaction coordinate. This corresponds to the increase of the acidic strengt of the reactionary system at the initiation moment up to $pKa \approx -17$ (for C_3H_7 pKa=-15, for C_2H_4 - -11) according to correlational dependence $pKa = f(q_{H+})$ [4]. Heat effect of the reaction E_{int} decreases in the sequence of olefins C_4H_8, C_3H_6, C_2H_4. At the same time the activation energies E_a of the reaction of initiaton of electrophilic process of polymerization of olefins increases in the same sequence.

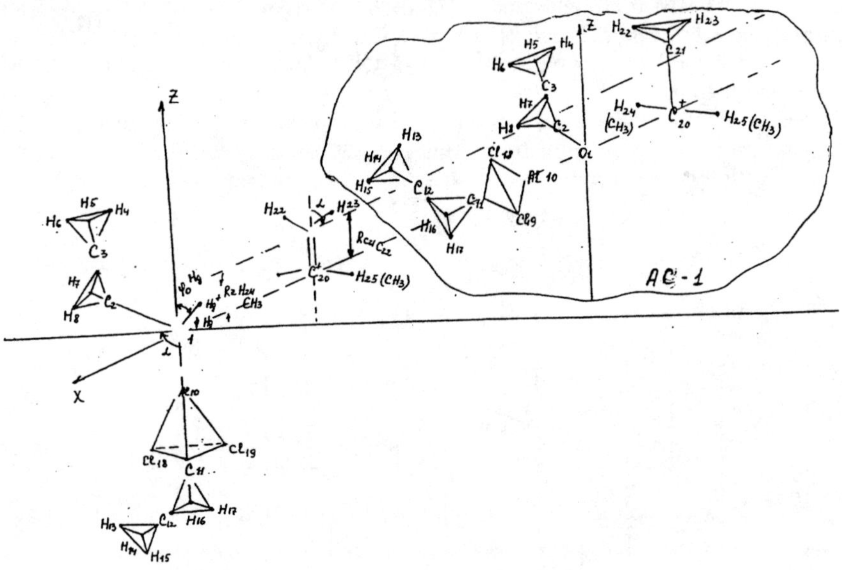

Figure 4.24. The mechanism of interactions of $C_2H_5AlCl_2 \cdot OHC_2H_5$ complex with olefins at performing proton-donor activity.

Table 4.28. Quantum-chemical interaction sof $C_2H_5AlCl_2 \cdot C_2H_5OH$ complex with olefins.

No.	Model	Activity	CNDO/2						MINDO/3
			φ_o, H9 grade	φ_o, C2 grade	$q_{H_9^+}$	$q_{C_2^+}$	Eint kJ mole	Ea kJ mole	Ea kJ mole
1.	$C_2H_5AlCl_2 \cdot C_2H_5OH \cdot C_2H_4$	H^+	35	–	+0.27	–	+546	-102	-29
2.	$C_2H_5AlCl_2 \cdot C_2H_5OH \cdot C_3H_6$	H^+	25	–	+0.29	–	+568	-74	-16
3.	$C_2H_5AlCl_2 \cdot C_2H_5OH \cdot C_4H_8$	H^+	15	–	+0.30	–	+597	-51	-13
4.	$C_2H_5AlCl_2 \cdot C_2H_5OH \cdot C_2F_4$	–	–	–	+0.24	–	-32	–	–
1.	$C_2H_5AlCl_2 \cdot C_2H_5OH \cdot C_2H_4$	R^+	–	40	–	+0.24	+550	-124	-37
2.	$C_2H_5AlCl_2 \cdot C_2H_5OH \cdot C_3H_6$	R^+	–	30	–	+0.24	+581	-86	-19
3.	$C_2H_5AlCl_2 \cdot C_2H_5OH \cdot C_4H_8$	R^+	–	20	–	+0.24	+613	-63	-15
4.	$C_2H_5AlCl_2 \cdot C_2H_5OH \cdot C_2F_4$	–	–	–	–	+0.27	-27	–	–
						+0.23			

It is known [55, 60], that CNDO/2 method at the calculation of energetic characteristics may overestimate their values by 3-5 times. That is why for the comparison and for additional control of the results obtained by CNDO/2 method it was performed the calculation of E_a values by MINDO/3, which reproduces the heat formation and the bond length between atoms. Absolute values of E_a in these methods differ from each other. However, tendencies in unitypical sequences of olefins to E_a increase are confirmed by both methods. This is in agreement with the experimental data on the activity of electrophilic polymerization of the studied monomers [2, 3].

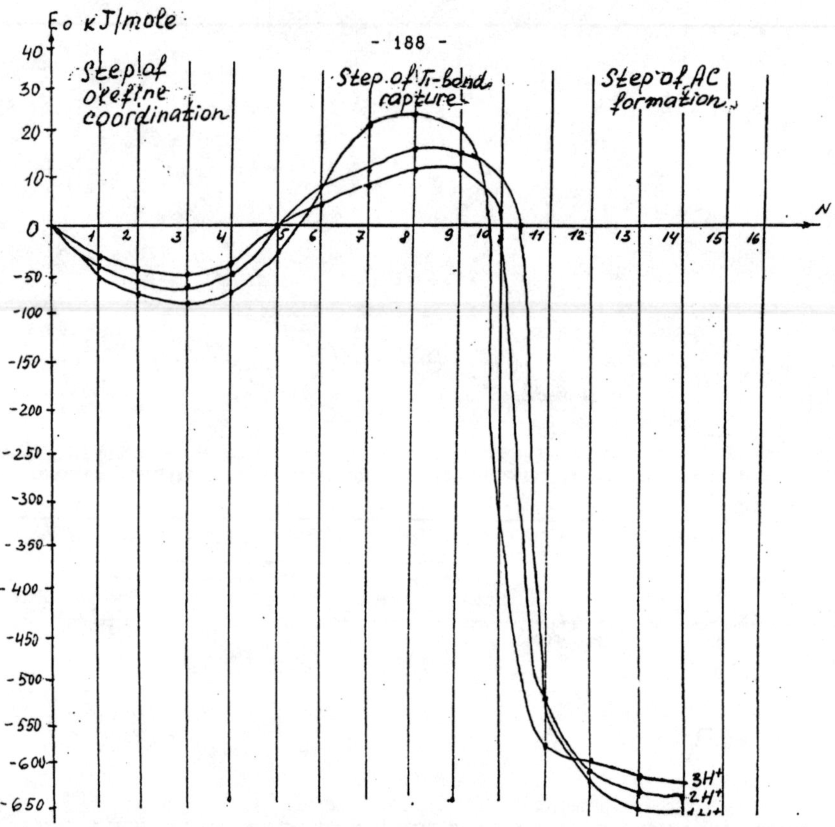

Figure 4.25. The change of total energy of the system on the reaction trajectory of the complex of $C_2H_5AlCl_2 \cdot C_2H_5OH$ with olefins at performing proton-donor activity (CNDO/2 method): 1. $C_2H_5AlCl_2 \cdot C_2H_5OH \cdot C_4H_9$; 2. $C_2H_5AlCl_2 \cdot C_2H_5OH \cdot C_3H_7$; 3. $C_2H_5AlCl_2 \cdot C_2H_5OH \cdot C_2H_4$.

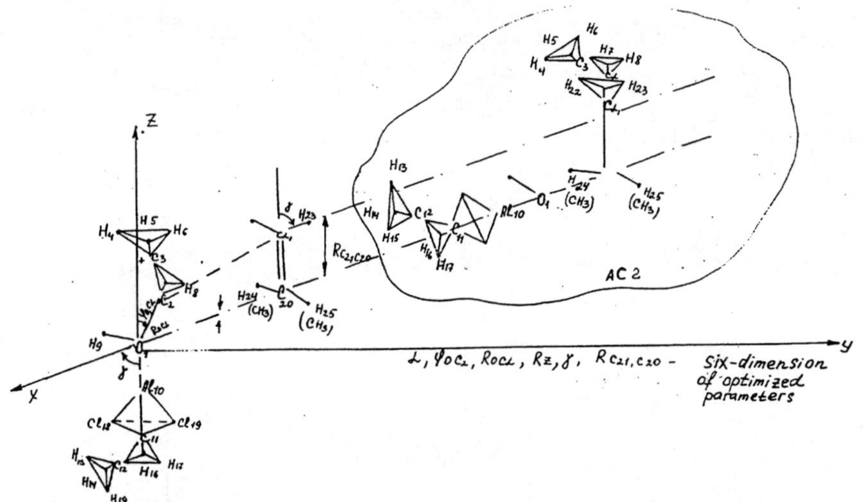

Figure 4.26. The mechanism interaction of the complex of $C_2H_5AlCl_2 \cdot C_2H_5OH$ with olefins at performing carbcationic activity.

The change of total energy of the systems - E_0 at the interaction of $C_2H_5AlCl_2 \cdot C_2H_5OH$ complex with ethylene, propylene and isobutylene by the reaction coordinate – $R_{0_1 0_2 0}$ - at the display of proton-donor activity by the complex is shown in the Table 4.25 (I direction of the reaction according to the scheme IV). The values of heat coefficients of the reaction E_{int} and activation energy E_a were determined from the curves. The lowest value of E_a is 51 kJ/mole (CNDO/2), it is characteristic for interaction of $C_2H_5AlCl_2$ OHC_2H_5 with isobutylene and is close to $E_a \approx 50$ kJ/mole (CNDO/2) at the initiation of isobutylene under the influence of $AlCl_3$ H_2O aquacomplex [110]. The comparison of the behavior of $C_2H_5AlCl_2 \cdot C_2H_5OH$ complex fragments at cationic polymerization of isobutylene shows, that in both cases similar coordination mechanisms of the initiation of the processes take place. This was also found for polymerization of propylene and ethylene in presence of the same systems (Figure 4.26).

Coordination mechanism of the interaction of $C_2H_5AlCl_2 \cdot C_2H_5OH$ complex with olefins at the display of carbcationic activity by the complex (II direction of the reaction, the scheme IV) is presented on the Figure 4.26. It shows geometric parameters of optimization of the total energy of the system. Energetic profile of this reaction is presented on the Figure 4.27 ($R_{0_1 C_{20}^+}$ - the reaction coordinate). The comparison of energetic parameters E_a and E_{int} at the display of R^+ and H^+-activities at the initiation of electrophilic polymerization of olefins shows, that the character of the change of these parameters and unitypical sequences is similar (Table 4.28).

The structure of AC of cationic polymerization of olefins (ethylene, propylene, isobutylene) on the basis of $C_2H_5AlCl_2 \cdot CH_5OH$ complex at the simultaneous display of proton-donor and carbcationic activity by ethyl alcohol is presented on the Figure 4.24 (AC1) and on the Figure 4.27 (AC2). These are polarized structures ($C_{20}^+ - 0_1$ is higher than covalent one), which may be considered as the border case of the ionic pair. AC1 (I direction of the reaction) possesses the following composition in dependence on the nature of polymerizing olefins:

1. $[C_2H_5AlCl_2 \cdot OC_2H_5]^{-\delta} \ldots C^{+\delta}H_2CH_3$ – for ethylene;

2. $[C_2H_5AlCl_2 \cdot OC_2H_5]^{-\delta} \ldots C^{+\delta}H(CH_3)_2$ – for propylene;

3. $[C_2H_5AlCl_2 \cdot OC_2H_5]^{-\delta} \ldots C^{+\delta}(CH_3)_3$ – for isobutylene.

 AC2 (II) direction of the reaction) is, respectively:

1. $[C_2H_5AlCl_2 \cdot OH]^{-\delta} \ldots C^{+\delta}H_2CH_2(CH_3)_2$ – for ethylene;

2. $[C_2H_5AlCl_2 \cdot OH]^{-\delta} \ldots C^{+\delta}HCH_3CH(CH_3)_2$ – for propylene;

3. $[C_2H_5AlCl_2 \cdot OH]^{-\delta} \ldots C^{+\delta}(CH_3)_2CH_2(CH_3)_2$ – for isobutylene.

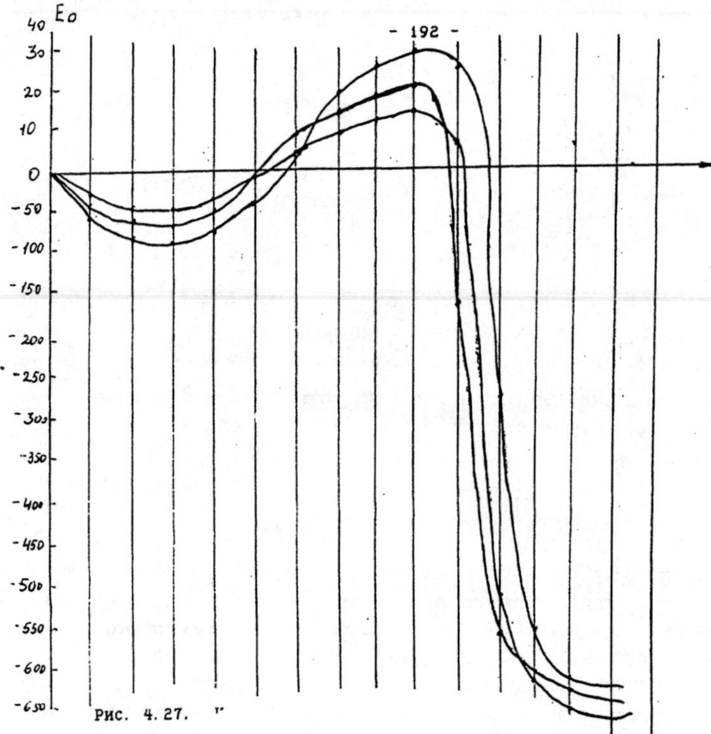

Figure 4.27. The change of total energy of the system on the reaction trajectory of the complex of $C_2H_5AlCl_2 \cdot C_2H_5OH$ with olefin at performing carbcationic activity (CNDO/2). 1. $C_2H_5AlCl_2 \cdot C_2H_5OH \cdot C_4H_9$; 2. $C_2H_5AlCl_2 \cdot C_2H_5OH \cdot C_3H_7$; 3. $C_2H_5AlCl_2 \cdot C_2H_5OH \cdot C_3H_7$.

 AC1 and AC2 differ sufficiently from each other by the composition by both nature of counterions and carbcations. Nevertheless, quantum-chemical parameters (E_a and E_{int}, Table (4.28), which reflect the energetics of AC formation, point out the possibility of simultaneous existence of the active centers of two different types – AC1 and AC2, at the initiation of electrophilic polymerization of olefins in presence of

$C_2H_5AlCl_2 \cdot C_2H_5OH$ complex. This correlates with the experimental resul [26, 123], which testify the reactions proceeding in parallel according to the ways I and II (scheme IV). The data of IR-spectra of products of oligomerization of $C_8^=$ in presence of $C_2H_5AlCl_2$-CD_3OH, UV and PMR-spectra (for $C_2H_5AlCl_2 \cdot C_6H_5CH_2OH$), GLC (for $C_2H_5AlCl_2 \cdot C_4H_9OH$), comparison of the values of MM of products, the degree of their unsaturation and content of functional groups [26] allow to make a conclusion that the ratio AC1/AC2 in the reactionary system will be defined by the stability of carbcations of alcohol as cocatalyst, which depends on nature of olefin, and counterion and other stabilizing factors, which are responsible for the predominant direction of the reaction by the end.

Consequently, independent on chemical structure of the olefin (ethylene, propylene, isobutylene) at their polymerization in presence of $C_2H_5AlCl_2 \cdot ROH$ complexes ethyl alcohol performs both proton-donor and carbcationic activity. In this case, as quantum-chemical calculations show, the active centers of two types are formed AC1 and AC2. The possibility of their formation is equally probable. Acidic-catalytic properties of $C_2H_5AlCl_2 \cdot C_2H_5OH$ are performed only at the direct participation of substrates $(C_2H_4, C_3H_6$ or $C_4H_8)$, similarly to aquacomplexes of aluminium chlorides [112].

4.8. Complexes of alcohols with magnesium chlorides.

Magnesium chlorides form donor-acceptor complexes at the interaction with alcohols. Quantum-chemical data of complexes $R_nMgCl_{2-n} \cdot ROH$ are presented in the Table 4.29, characterized by high negative values of total energy of the system as the guarantee of their stability and sufficient energies of the complex formation (E_c=-269±40 kJ/mole). In this case the part of electron charge - Δq, transferred from donor to hydrogen atom of the acceptor (ROH) was equal 0.05÷0.02. The transition of nonlinked electrons of oxygen in the molecule of alcohol is performed from p_y-orbitals to vacant Sp^2-orbitals of Mg^{2+}. The overlapping of d-orbitals of one of Cl atoms and p_z orbital of the oxygen atom is also sufficient (corresponding overlapping integral equals 0.18). MOdels of complexes of alcohols with magnesium halogenides are shown on the Figure 4.28. All complexes of alcohols with R_nMgCl_{2-n} are characterized by low Brenstedt acidity (pKa=+7.1÷+12.0) (Table 4.29) with no dependence on nature of ligand surrounding of Mg. The increase of pKa of $R_nMgCl_{2-n} \cdot ROH$ complexes in the sequence of alcohols: CH_3OH, C_2H_5OH, C_4H_9OH – is stipulated by their nature. But it is impossible to explain acidic-catalytic properties of $R'_nMgCl_{2-n} \cdot ROH$, $RMgCl \cdot ROH$ and $Cl_2Mg \cdot ROH$ complexes (Figure 4.29) by the acidic strength only, as well as in the case of $R_nMgCl_{2-n} \cdot H_2O$ aquacomplexes. Geometry and electron structure of these complexes points out the singularity of the coordination of the alcohol molecule according to $MgCl_2$ or $RMgCl$. The molecules $R_2Mg \cdot ROH$ complex (Figure 4.29), i.e. in the absence of completely occupied d-jackets, possess usual coordination of the alcohol molecule, and it is also observed uniformal removal o the oxygen atom from R ligands of Mg heteroatom in ROH molecule [55]. Apparently, singularity of the coordination of the alcohol molecule in $RMgCl \cdot ROH$ and $Cl_2Mg \cdot ROH$ complexes (comparing with the models of $Cl_3Al \cdot ROH$ complex [123], for example) is characteristic for double-valent atoms only, for Mg in particular. Such unusual coordination of alcohol in complexes with halogenides of magnesium (Figure 4.28) increases their sizes. In this connection it should be expected the formation of counterions of $Cl_2Mg \cdot ROH$ and $RMgCl \cdot ROH$ complexes of big sizes at the initiation of electrophilic processes. The corresponding counterions $[Cl_2Mg \cdot OH]^{-\delta}$ and $[RMgCl \cdot OH]^{-\delta}$ possess the radii in the range of $0.7 \div 0.9$ nm comparing with the sizes of $[AlCl_3 \cdot OH]^{-\delta}$ counterion for example, which ionic radius equals 0.38 nm [104].

Table 4.29. Quantum-chemical characteristics of $R_nMgCl_{2-n} \cdot OH$ complexes (CNDO/2).

No.	Complex	q_{H^+}	pKa	P_{OH}	P_{OC}	Activity	No. of group
1.	$MgCl_2 \cdot OHCH_3$	+0.20	+7.5	0.93	0.99	H^+	
2.	$MgCl_2 \cdot OHC_2H_5$	+0.20	+7.5	0.93	0.93	H^+, R^+	I
3.	$MgCl_2 \cdot OHC_3H_7$	+0.19	+7.7	0.93	0.90	R^+	
4.	$MgCl_2 \cdot OHC_4H_9$	+0.17	+12.0	0.94	0.89	R^+	
5.	$MgClCH_3 \cdot OHCH_3$	+0.20	+7.5	0.94	0.98	H^+	
6.	$CH_3ClMg \cdot OHC_2H_5$	+0.19	+7.7	0.94	0.94	H^+, R^+	II
7.	$CH_3ClMg \cdot OHC_3H_7$	+0.19	+7.7	0.94	0.91	R^+	
8.	$CH_3ClMg \cdot OHC_4H_9$	+0.17	+12.0	0.93	0.90	R^+	
9.	$(CH_3)_2Mg \cdot OHCH_3$	+0.20	+7.5	0.96	0.97	H^+, R^+	
10.	$(CH_3)_2Mg \cdot OHC_2H_5$	+0.19	+7.7	0.94	0.95	H^+, R^+	III
11.	$(CH_3)_2Mg \cdot OHC_3H_7$	+0.18	+8.5	0.94	0.91	R^+	
12.	$(CH_3)_2Mg \cdot OHC_4H_9$	+0.18	+8.5	0.93	0.90	R^+	
13.	$C_2H_5ClMg \cdot OHCH_3$	+0.20	+7.5	0.93	0.99	H^+	
14.	$C_2H_5ClMg \cdot OHC_2H_5$	+0.19	+7.7	0.94	0.93	H^+, R^+	IV
15.	$C_2H_5ClMg \cdot OHC_3H_7$	+0.18	+8.5	0.93	0.90	R^+	
16.	$C_2H_5ClMg \cdot OHC_4H_9$	+0.17	+12.0	0.93	0.90	R^+	
17.	$(C_2H_5)_2 \cdot Mg \cdot OHCH_3$	+0.20	+7.5	0.93	1.00	H^+	
18.	$(C_2H_5)_2Mg \cdot OHC_2H_5$	+0.19	+7.7	0.93	0.93	H^+, R^+	V
19	$(C_2H_5)_2Mg \cdot OHC_3H_7$	+0.18	+8.5	0.94	0.91	R^+	
20.	$(C_2H_5)_2Mg \cdot OHC_4H_9$	+0.17	+12.0	0.93	0.90	R^+	
21.	$C_3H_7ClMg \cdot OHCH_3$	+0.19	+7.7	0.92	0.98	H^+	
22.	$C_3H_7ClMg \cdot OHC_2H_5$	+0.18	+8.5	0.93	0.94	H^+, R^+	VI
23.	$C_3H_7ClMg \cdot OHC_3H_7$	+0.18	+8.5	0.93	0.90	R^+	
24.	$C_3H_7ClMg \cdot OHC_4H_9$	+0.17	+12.0	0.93	0.90	R^+	
25.	$(C_3H_7)_2Mg \cdot OHCH_3$	+0.20	+7.5	0.93	0.99	H^+	
26.	$(C_3H_7)_2Mg \cdot OHC_2H_5$	+0.19	+7.7	0.94	0.94	H^+, R^+	VII
27.	$(C_3H_7)_2Mg \cdot OHC_3H_7$	+0.17	+12.0	0.93	0.92	H^+, R^+	
28.	$(C_3H_7)_2Mg \cdot OHC_4H_9$	+0.17	+12.0	0.94	0.89	R^+	
29.	$C_4H_9ClMg \cdot OHCH_3$	+0.20	+7.5	0.94	0.99	H^+	
30.	$C_4H_9CLMg \cdot OHC_2H_5$	+0.20	+7.5	0.94	0.93	H^+, R^+	VIII
31.	$C_4H_9ClMg \cdot OHC_3H_7$	+0.19	+7.7	0.94	0.90	R^+	
32.	$C_4H_9ClMg \cdot OHC_4H_9$	+0.18	+8.5	0.94	0.89	R^+	
33.	$(C_4H_9)_2Mg \cdot OHCH_3$	+0.20	+7.5	0.92	0.98	H^+	
34.	$(C_4H_9)_2Mg \cdot OHC_2H_5$	+0.20	+7.5	0.94	0.95	H^+, R^+	IX
35.	$(C_4H_9)_2Mg \cdot OHC_3H_7$	+0.18	+8.5	0.94	0.91	R^+	
36.	$(C_4H_9)_2Mg \cdot OHC_4H_9$	+0.17	+12.0	0.93	0.90	R^+	

Analyzing the values of the bonds orders of alcohol molecule P_{OH} and P_{OC} for $R_nMgCl_{2-n}\cdot ROH$ complexes, it may be easily noticed that in the case of application of methyl alcohol proton-donor activity (H^+) is preferable, and isopropyl and tret-butyl alcohols possess carbcationic activity (R^+).

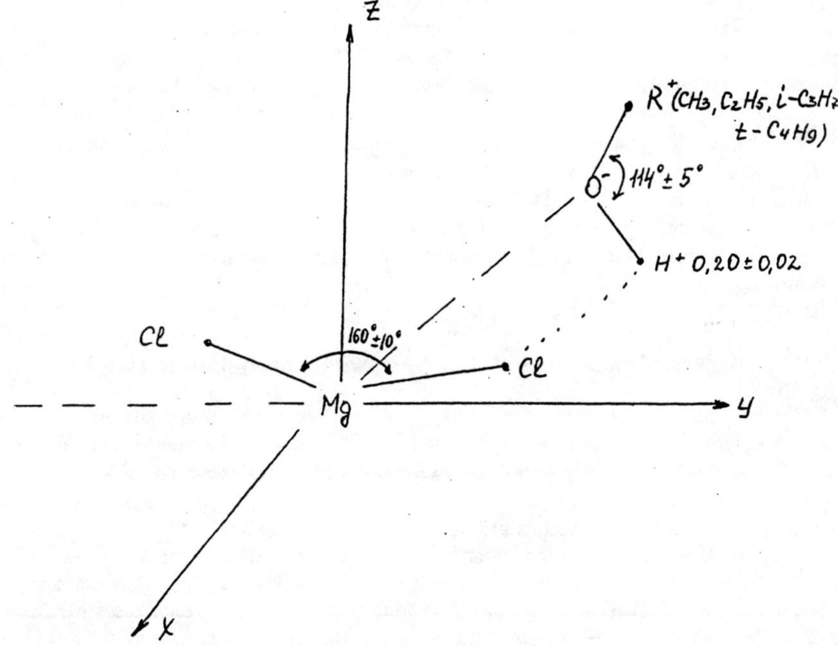

Figure 4.28. The models of complexes of $Cl_3Mg\cdot ROH$ (CNDO/2).

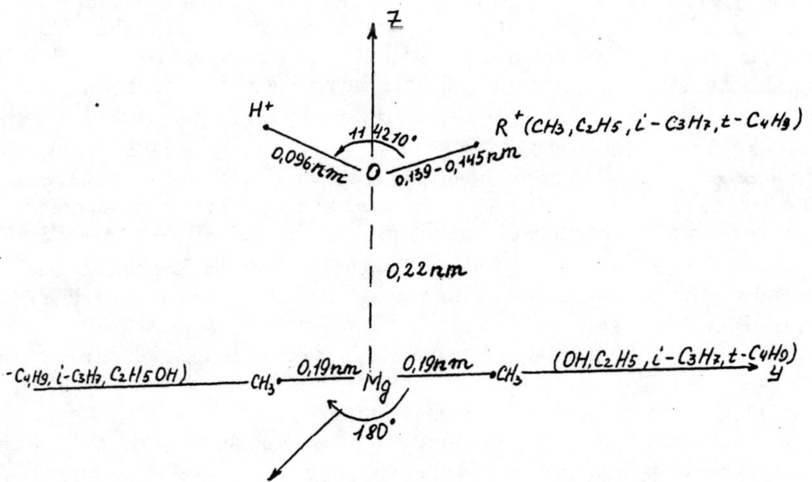

Figure 4.29. The models of complexes of $R_nMgCl_{2-n}\cdot ROH$ (CNDO/2).

$R_nMgCl_{2-n} \cdot OHC_2H_5$ complexes possess equal probabilities for alcohol to display R^+ or H^+ activity. Models of $R_nMgCl_{2-n} \cdot ROH$ complexes, which possess the nature of the ligand surrounding of heteroatom (Al, Mg) coinciding with that of the alcohol radical, i.e. the complexes $(CH_3)_2Mg \cdot OHCH_3$, $(C_2H_5)_2Mg \cdot OHC_2H_5$, $(C_3H_7)_2Mg \cdot OHC_3H_7$, $(C_4H_9)_2Mg \cdot OHC_4H_9$, show also equal probability for alcohols to display H^+ or R^+ activity in these complexes, which is similar to $R_nAlCl_{3-n} \cdot ROH$ complexes. As the authors imagine, one of possible causes explaining this fact is the coincidence of the symmetry groups of alcohol and ligands of Mg(Al).

Concluding the analysis of the data of quantum-chemical calculations of $R_nMgCl_{2-n} \cdot ROH$ complexes, let us point out, that independent on the nature of Mg compound the acidic strength of alcohols - magnesium chlorides complexes (acting as Brenstedt acids) is low. As in the case of $R_nAlCl_{3-n} \cdot ROH$ complexes, this supposes the direct participation of the reacting bases, olefins in particular, in the formation of AC. It also supposes the coordinated mechanisms of initiation and chain growth at cationic polymerization.

4.9. Complexes of alcohols with boron fluorides.

Boron fluorides possess the ability to form donor-acceptor complexes with alcohols [9]. Parameters of the acidic strength, P_{OH} and P_{OC} bonds orders and the character of the preferable activity of alcohol in $R_nBF_{3-n} \cdot ROH$ complexes are presented in the Table 4.30. For more comfort in the analysis and comparison of the results of quantum-chemical calculations the models are subdivided into 13 groups, differing by chemical nature of the ligand surrounding of boron. Electron structures of complexes before the optimization of their geometry according to the method of variable metrics are presented on the Figure 4.30. They represent tetrahedral or distorted tetrahedral configurations with the following parameters: the lengths of bonds $r_{OH}=0.11$ nm, $r_{BF}=0.134$ nm, $r_{OB}=0.17\pm0.005$ nm, $r_{CH}=0.109$ nm, $r_{CC}=0.150\pm0.004$ nm, the angle between ligands of B equals $(109\pm18)°$, valent angle of the alcohol $<HOR=(114\pm10)°$. The calculation gave high negative values of the total energy of the system ($E_0=-320590 \div -370990$ kJ/mole), which characterize the stability of complexes. The transition of nonlinked electrons of the oxygen in the alcohol molecule is performed from p_y-orbital to vacant Sp^2-orbitals of B^{3+}. The charge on hydrogen atom H^+ in the alcohol molecule (q_{H+}) in $R_nBF_{3-n} \cdot ROH$ complexes possesses the tendency to the decrease from +0.24 down to +0.19 at the increase of the branching of R radical in the alcohol molecule in all 13 groups being studied (Table 4.30). In accordance with the correlational ratio $q_{H+} \sim f(pKa)$ the acidic strength of the complexes decreases from $+0.5 \div +1.5$ to $+7.3 \div +7.7$ in the sequence of alcohols CH_3OH, C_2H_5OH, C_3H_7OH, C_4H_9OH. The bond order P_{OC} possesses the inclination to decrease with the decrease of the acidic strength of the complexes. P_{OH} of all 13 groups of $R_nBF_{3-n} \cdot ROH$ complexes does not practically change and equals $0.92 \div 0.94$ (the accuracy range of CNDO/ method at the estimation of the bond orders is 0.02). The comparison of the values of P_{OH} and P_{OC} exhibits, that ethyl and methyl alcohols possess the predominant proton-donor activity (H^+) in the complexes with boron fluorides, and tret-butyl one – carbcationic activity (R^+). In the case of application of isopropyl alcohol the possibility of the display of H^+ and R^+ activities is equally probable. Two alternative possibilities of fragmentation of the alcohol molecules in dependence on the structure of R radical is proved experimentally

on the example of Cl3Al·ROH complex [26]. Quantum-chemical calculations testify the fact that similar possibilities are characteristic for ROH fragmentation also in the complexes with boronorganic compounds R_nBF_{3-n}. Moreover, it was observed that R_nBF_{3-n}·ROH complexes possess equal probability of the display of H^+ and R^+ activities in the case if the nature of the ligand surrounding of boron is similar to the nature of the alcohol radical, i.e. in the case of complexes $(CH_3)_3B$·$OHCH_3$, $(C_2H_5)_3B$·OHC_3H_5, $(C_4H_9)_3B$·OHC_4H_9. One of possible causes explining this result (as in the case of R_nAlCl_{3-n}·ROH and R_nMgCl_{3-n}·ROH complexes) is the coincidence of the symmetry groups of the alcohol radicals and the radicals of the ligand surrounding of heteroatoms. However this supposition requires experimental verification.

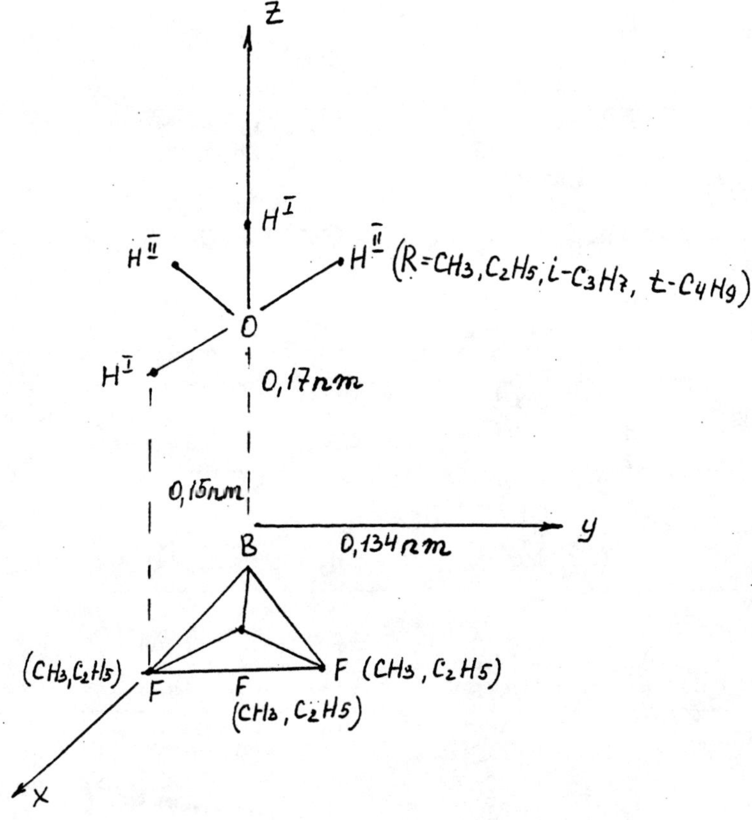

Figure 4.30. Electron structure and geometry of the models of aquacomplexes of boron fluorides (R_{BC}=0.156-0.164 nm) (CNDO/2).

Table 4.30. Parameters of the acidic strength and the bond orders P_{OH} and P_{OC} for the models of $R_nBF_{3-n} \cdot ROH$ complexes, $R=CH_3, C_2H_5, C_3H_7, C_4H_9$ (CNDO/2).

No.	Complex model	q_{H^+}	pKa	P_{OH}	P_{OC}	Activity	No. of group
1.	$F_3B \cdot OHCH_3$	+0.24	+1.0	0.92	0.98	H^+	
2.	$F_3B \cdot OHEt$	+0.23	+1.5	0.92	0.97	H^+	I
3.	$F_3B \cdot OHC_3H_7$	+0.21	+7.1	0.92	0.94	H^+, R^+	
4.	$F_3B \cdot OHC_4H_9$	+0.20	+7.3	0.93	0.90	R^+	
5.	$F_2CH_3B \cdot OHCH_3$	+0.23	+1.5	0.92	1.01	H^+	
6.	$F_2CH_3B \cdot OHEt$	+0.22	+4.4	0.92	0.96	H^+	II
7.	$F_2CH_3B \cdot OHC_3H_7$	+0.20	+7.3	0.93	0.90	R^+	
8.	$F_2CH_3B \cdot OHC_4H_9$	+0.19	+7.7	0.93	0.89	R^+	
9.	$F(CH_3)_2 \cdot OHCH_3$	+0.23	+1.5	0.92	1.0	H^+	
10.	$F(CH_3)_2B \cdot OHEt$	+0.21	+7.1	0.94	0.97	H^+	III
11.	$F(CH_3)_2B \cdot OHC_3H_7$	+0.20	+7.3	0.92	0.94	H^+, R^+	
12.	$F(CH_3)_2B \cdot OHC_4H_9$	+0.20	+7/3	0.93	0.90	R^+	
13.	$(CH_3)_3B \cdot OHCH_3$	+0.24	+1.0	0.94	0.96	H^+, R^+	
14.	$(CH_3)_3B \cdot OHEt$	+0.23	+1.5	0.92	0.95	H^+	IV
15.	$(CH_3)_3B \cdot OHC_3H_7$	+0.22	+4.4	0.94	0.91	R^+	
16.	$(CH_3)_3B \cdot OHC_4H_9$	+0.20	+7.3	0.92	0.89	R^+	
17.	$F_2EtB \cdot OHCH_3$	+0.23	+1.5	0.93	0.98	H^+	
18.	$F_2EtB \cdot OHEt$	+0.22	+4.4	0.92	0.96	H^+	V
19.	$F_2EtB \cdot OHC_3H_7$	+0.22	+4.4	0.94	0.93	H^+, R^+	
20.	$F_2EtB \cdot OHC_4H_9$	+0.21	+7.1	0.93	0.90	R^+	
21.	$F(Et)_2B \cdot OHCH_3$	+0.24	+1.0	0.92	0.97	H^+	
22.	$F(Et)_2B \cdot OHEt$	+0.23	+1.5	0.92	0.95	H^+	VI
23.	$F(Et)_2B \cdot OHC_3H_7$	+0.21	+7.1	0.92	0.93	H^+, R^+	
24.	$F(Et)_2B \cdot OHC_4H_9$	+0.20	+7.3	0.93	0.90	R^+	
25.	$(Et)_3B \cdot OHCH_3$	+0.22	+4.4	0.92	0.90	H^+	
26.	$(Et)_3B \cdot OHEt$	+0.21	+7.1	0.93	0.95	H^+, R^+	VII
27.	$(Et)_3B \cdot OHC_3H_7$	+0.21	+7.1	0.92	0.93	H^+, R^+	
28.	$(Et)_3B \cdot OHC_4H_9$	+0.20	+7.3	0.93	0.90	R^+	
29.	$F_2C_3H_7B \cdot OHCH_3$	+0.23	+1.5	0.92	0.98	H^+	
30.	$F_2C_3H_7B \cdot OHEt$	+0.22	+4.4	0.3	0.97	H^+	VIII
31.	$F_2C_3H_7B \cdot OHC_3H_7$	+0.20	+7.3	0.92	0.93	H^+, R^+	
32.	$F_2C_3H_7B \cdot OHC_4H_9$	+0.20	+7.3	0.92	0.89	R^+	
33.	$F(C_3H_7)_2B \cdot OHCH_3$	+0.24	+1.0	0.92	0.99	H^+	
34.	$F(C_3H_7)_2B \cdot OHEt$	+0.24	+1.0	0.93	0.96	H^+	IX
35.	$F(C_3H_7)_2B \cdot OHC_3H_7$	+0.21	+7.3	0.93	0.94	H^+, R^+	
36.	$F(C_3H_7)_2B \cdot OHC_4H_9$	+0.20	+7.3	0.92	0.89	R^+	
37.	$(C_3H_7)_3B \cdot OHCH_3$	+0.22	+4.4	0.92	0.99	H^+	
38.	$(C_3H_7)_3B \cdot OHEt$	+0.22	+4.4	0.92	0.96	H^+	X
39.	$(C_3H_7)_3B \cdot OHC_3H_7$	+0.21	+7.1	0.92	0.93	H^+, R^+	
40.	$(C_3H_7)_3B \cdot OHC_4H_9$	+0.19	+7.7	0.92	0.89	R^+	

41.	$F_2C_4H_9B \cdot OHCH_3$	+0.23	+1.5	0.92	0.99	H^+	
42.	$F_2C_4H_9B \cdot OHEt$	+0.21	+7.1	0.93	0.96	H^+	XI
43.	$F_2C_4H_9B \cdot OHC_3H_7$	+0.21	+7.1	0.92	0.93	H^+, R^+	
44.	$F_2C_4H_9B \cdot OHC_4H_7$	+0.20	+7.3	0.93	0.89	R^+	
45.	$F(C_4H_9)_2B \cdot OHCH_3$	+0.23	+1.5	0.92	1.00	H^+	
46.	$F(C_4H_9)_2B \cdot OHEt$	+0.22	+4.4	0.93	0.97	H^+	XII
47.	$F(C_4H_9)_2B \cdot OHC_3H_7$	+0.22	+4.4	0.93	0.94	H^+, R^+	
48.	$F(C_4H_9)_2B \cdot OHC_4H_9$	+0.19	+7.7	0.93	0.90	R^+	
49.	$(C_4H_9)_3B \cdot OHCH_3$	+0.22	+4.4	0.93	1.00	H^+	
50.	$(C_4H_9)_3B \cdot OHEt$	+0.21	+7.1	0.93	0.96	H^+	XIII
51.	$(C_4H_9)_3B \cdot OHC_3H_7$	+0.20	+7.3	0.93	0.93	H^+, R^+	
52.	$(C_4H_9)_3B \cdot OHC_4H_9$	+0.19	+7.7	0.93	0.91	H^+, R^+	

Consequently, according to the data of quantum-chemical calculations the complexes are characterized by low acidic strength (+0.5<pKa<+7.7) with no dependence on the nature of boron ligands. The possibility for alcohols in the complexes with boron fluorides to display both proton-donor and carbcationic activity in the reactions of cationic polymerization of olefins is stipulated by the chemical structure of the radicls R of the alcohol. In the end the display of acidic-catalytic properties is supposed to be performed only in presence of a monomer because of low acidic strength of $R_nBF_{3-n} \cdot ROH$ complexes (similar to $R_nAlCl_{3-n} \cdot ROH$ and $R_nMgCl_{3-n} \cdot ROH$ ones). In this case mechanisms of the stages of initiation and chain growth at the polymerization of isobutylene (propylene, ethylene) are also similar to the mechanisms, which are realized in presence of the complexes of alcohols with aluminium chlorides and magnesium chlorides.

4.10. Correlation of proton-donor and carbcationic activity of alcohols of the complexes of Lewis and Brenstedt acids.

The calculation of complexes of alcohols (CH_3OH, C_2H_5OH, i-C_3H_7, t-C_4H_9OH) with different Lewis acids (R_nAlCl_{3-n}, R_nBF_{3-n}, R_nMgCl_{3-n}) points out the dependence of the direction of the alcohol fragmentation on the nature of its radical and the acid type. The experimental material existing in literature on cationic properties of the systems "alcohol - acidic agent" may be also classified according to the character of the alcohol behavior, i.e. by preferable display of protonic or carbcationic activity by it, or simultaneous display of both activities. The comparison of calculated and experimental data allow to control the correctness of the set notions about the ways of the alcohols fragmentation (Table 4.31).

One of the main factors, defining the direction of the fragmentation, is the stability of forming ionic particles. The presence of substitutors at α-atom of carbon and the possibility to link hydroxyl groups into stable anion leads to the appearance of reliably identified carbonium ions, for example in the system $(C_6H_5)_3COH-H_2SO_4$ [8, p. 15]. The existence of stable carbcations, connected with high stabilty (low nucleophilicity) of counterions was obtained by NMR method in the case of the complex formation of alcohol with "super"-acid [3]:

$$(CH_3)_3C - \int - OH \xrightarrow{SbF_5FSO_3H} (CH_3)_3C^+SbF_5FSO_3 + H_2).$$

The formation of $(CH_3)C^+BF_4^-$ was also observed at the interaction of tret-butyl alcohol with another strong acidic system $BF_3 \cdot HF$ [3]. Thus, tret-substituted alcohols endure the break of C-O bond in presence of strong acids.

If the conditions of the experiment provide no stability of carbonium ions, the direction of fragmentation of the substance complex-linked with Lewis acid is defined by other factors, in particular by chemical linking of the detached group into stable compounds. For example, gas-phase condensation of tret-butyl cations with CH_3OH, C_2H_5OH in presence of the proton acceptor - ammonia gas - proceeds according to the following scheme:

$$\text{tret} - C_4H_9 + OR - \int -H \xrightarrow{\ NH_3\ } \text{tret} - C_4H_9OR^- + NH_4^+,$$

i.e. through the break of O-H bond and not C-O one, as it follows from the calculation.

In analog, the break of growing carbcations of polyisobutylene at the reaction with alcohol proceeds with the formation of ester compositions, i.e. through the break of O-H bond [124]. Evidently, this is supported by lower activity of carbonium ions in the composition of ionic pairs in comparison with free carbcations (calculation) and by the influence of counterion, promoting the regeneration of the catalyst in the act of the chain break or in the proton transfer to the monomer, accompanying the latter:

$$\sim C^+A^- + RO - \int -H \rightarrow \sim COR + H^+A^-$$

In the case of phenols application the fragmentation of ROH by O-H bond in conditions of cationic polymerization of isobutylene [124] is promoted by higher H-acidity, than aliphatic alochols possess, and pp=pp conjugation of the oxygen atom with aromatic nucleus, which makes the break of C(phenyl)-O bond difficult.

The approximately equal probability of alcohols fragmentation according to both above mentioned directions (RfO-H or R-OfH), which follows from the calculations of the complexes of Lewis acids (MeX_n) with alcohols, is confirmed by the experimental data. For example, at the application of nucleophilic acceptor of difficult polymerizing diisobutylene (C_8H_{16}) is was shown, that along with protonation of olefins there takes place its reaction with the alcohol carbcation [19]. In other words, the following reactions proceed in parallel:

$$\underset{R+}{\overset{Cl\ldots H}{\overset{\diagup\quad\diagdown}{C_2H_5AlCl\cdot O}}}+nC_8H_{16}\left[\begin{array}{l}I\quad H(C_8H_{16})_n^+C_2H_5AlCl_2OR^-\\[2ex]II\quad R(C_8H_{16})_n^+C_2H_5AlCl_2OH^-\end{array}\right.$$

The latter of the reactions pointed out is supported by the formation of H...Cl bonds (see the reaction), which makes the access of proton for olefin molecule difficult [19].

Two alternative possibilities of the alcohol fragmentation in the complexes $R_nMeCl_{3-n}\cdot R'OH$ in dependence on chemical structure of radical are persuasively presented for complexes with Lewis acid BF_3. In the case of CH_3OH and C_2H_5OH the absence of the possibility of carbonium ions stabilization by counterion or solvent defines the preferable proton-donor activity of the alcohol in the complex with BF_3 [125]. For adamantols-1-2 the main role in the transformatin of $BF_3\cdot2CH_3COOH$ complexes is played by the formation of adamantyl-cations, which are stabilized by means of the secondary reactions [52].

Consequently, according to the calculations and the experimental data fragmentation of alcohols, substitute at α-atom of carbon in complexes with strong acids (for example, with $(C_6H_5)COH\text{-}H_2SO_4$, $SbF_5FSO_3H\cdot BF_3\cdot HF$) proceeds by C-O bond with the formation of carbonium ions. In opposite, at the decrease of the acidic strength of the complex forming agent and especially at the application of alcohols with nonbranched radical (CH_3, C_2H_5, etc.) $R_nMeCl_{3-n}\cdot ROH$ complexes display preferably proton-donor properties. To add the calculated data, the experiment shows that the behavior of complex systems depends on specific interactions between components, the nature of anion and on other factors. In some case it is precisely these factors, which define the direction of the alcohol fragmentation according to the way I or II of the reaction.

Table 4.31. Systematization of the experimental and calculated data on proton-donor and carbcationic activity of alcohols in complexes.

No.	System	Experimental behavior of alcohol	Activity	Literature
1.	$FSO_3H\text{-}SbF_5+$ $t\text{-}C_4H_9OH$ $(t\text{-}C_5H_{11}OH,$ $t\text{-}C_7H_{15}OH)$	Stable cations are formed at -60°C, studied by PMR method: $(CH_3)_3COH\xrightarrow{SbF_5FSO_3H}$. $(CH_3)_3C^+SbF_5FSO_3^-+H_3O^+$ Solutions of third-degree and secondary alcohols in $SbF_5\text{-}SO_2$ or $SbF_5\text{-}SO_2ClF$ form carbonium ions.	R^+	Olah G.A. Friedel, Crafts Chemistry, New York-London-Sydney-Toronto, 1972, pp. 483-485.
2.	$HF+BF_3+t\text{-}C_4H_9OH$	$HF+BF_3+(CH_3)_3COH\rightarrow(CH_3)_3$ $C+BF_4^-$	R^+	p. 485, the same place

3.	$H_2SO_4 + (C_6H_5)_3OH$ 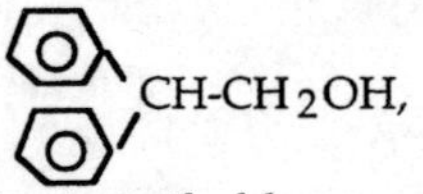 [structure: CH-CH$_2$OH diphenyl], heptamethyl-benzene alcohol, $\alpha\alpha,2,4,6$-pentamethylbenzene alcohol, $\alpha\alpha,2,4,6$-pentamethylbenzene alcohol $t\text{-}C_4H_9OH$.	Measurement of the change of freezing point displays the reaction proceeding: $(C_6H_5)_3COH + 2H_2OSO_4 \rightarrow (C_6H_5)_3C^+ + H_3O^+ + 2HSO_4^-$. It is pointed out by colour of solutions (from yellow to red $\lambda = 4000 \div 5000$Å. Absorption bend at 2930 Å points out the formation of $(CH_3)_3C^+$.	R^+	pp. 15-16, 38-39.
4.	$B^+H + Ph_3COH$ (Ph_2CHCH_2OH)	In the case of less nucleophilic alcohols the rates of the reactions proceeding with the participation of carbonium ions often depend on Hammet's acidity function. Carbonium ions appear resulting two-stage process: $ROH + B^+H \rightleftharpoons R^+OH_2 + B$ (fast) $R^+OH_2 \rightarrow R^+ + H_2O$ (slow) This mechanism is observed at the regroupping of pinacone into pinacoline and at racemisation of butanol-2.	R^+	The same paper, p. 53.
5.	$AlCl_3 + CH_3OH$	The complex alkylates benzene and isobutane through C^+H_3.	R^+	Olah G.A. Friedel, Crafts Chemistry, New York-London-Sydney-Toronto, 1972, p. 259
6.	$(C_6H_5)_3C^+$ alkylates 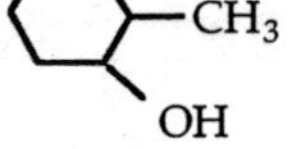 [structure: cyclohexene ring with CH$_3$ and OH]	Three-phenylcarbonyl alkylates o-crezole in sulfuric or acetic acid: 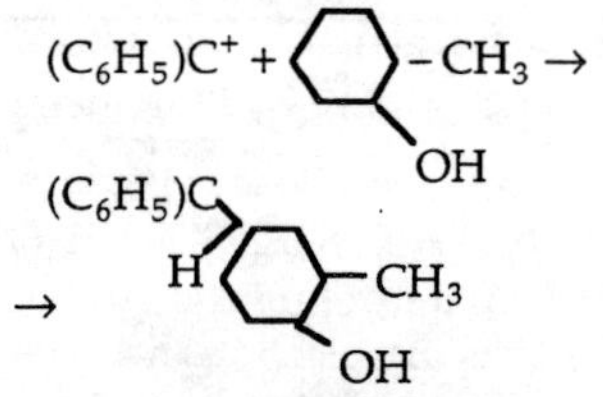 $(C_6H_5)C^+ +$ [ring with CH$_3$, OH] $\rightarrow$ $(C_6H_5)C$ [ring with H, CH$_3$, OH] $\rightarrow$ In analog it proceeds diphenylmethylation of aromatic compounds by diphenylcarbonyl.	R^+	Plesh P. Cationic Polymerization, Moscow, Mir, 1966, p. 21.

7.	$H^+ + C_2H_5OH$	Quantum-chemical CNDO/2 method shows in valent approximation that at ethyl alcohol protonation $O-R(C_2H_5)$ bond is weakened by the highest degree, which promoes dehydration of alcohol.	R^+	Chuvyilkin N.D., Jidomirov G.M., Kazansky V.V., "Kinetics and Catalysis", 1973, v. 14, p. 943 Chuvyilkin N.D., Jidomirov G.M., Kazansky V.V., "Kinetics and Catalysis", 1975, v. 16, p. 92.
8.	Alumosilicate catalyst $+C_2H_5OH$.	Method CNDO/BW. Maximal weakening of C-O bond (and of C_β-H one).	R^+	Segenya I.N., Miheikin G.M. Jidomirov g.M., Kazansky V.V., "Kinetics and Catalysis", 1980, v. 21, p. 1187
9.	a. ...+ second-C_4H_9OH b....+ROH $(R=C_6H_{13}, C_7H_{15}, C_8H_{17}, C_{12}H_{25})$. c....+$C_2H_5OH$ d....+ i-C_3H_7OH. e....+ i-C_3H_7OH.	Alkylation of aromatic hydrocarbons by alcohols: $...+second\text{-}C_4H_9OH \xrightarrow{BF_3}$ second-C_4H_9-... (yield 55.9%) $...+ROH \xrightarrow{BF_3BCCl} alkyl\text{-}$ benzenes $+H_2O$. $...+C_2H_5OH \xrightarrow{BF_3P_2O_5} ethyl\text{-}$ toluenes $+H_2O$ (yield 70%). $...+i\text{-}C_3H_7OH \xrightarrow{BF_3} cymoles$ $...+i\text{-}C_3H_7OH \xrightarrow{BF_3H_3PO_4}$ 1-ethyl-4-sopropylbutyl $+ H_2O$	R^+ R^+ R^+ R^+ R^+	Catalytic properties of substances (Ed. Reuter), Kiev, Naukova Dumka, 1968, p. 1462 The same. The same. The same. The same.
10.	Ni(Cu, Al) $+C_2H_5OH$.	Methods: Hukkel IR-spectroscopy (...). There were obtained absorption bends of C-O bond and C-C-O, CH-methylene groups and Al-O-C_2H_5.	H^+, R^+	Orlov A.N., Gagarin S., Kolgin G.K., Lyigin V.I. "Kinetics and Catalysis" 1973, v. 14, p. 1228.
11.	Solid acid Al_2O_3-$\gamma+OHCH_3$.	Millicen-Rudenberg method. $Al_2\tilde{O}_3$ surface was modeled by $Al\tilde{O}_3$, $Al\tilde{O}_4$, $Al\tilde{O}_5$ clusters. In the case of $Al\tilde{O}_4$ the bonds O-H and C-O are weakened, that promoes the formation of surface alcoholate and dehydration.	H^+, R^+	Gohbert P.Ja., Litinsky A.O., Shatkovsky D.B., Hardin A.P. "Conference on atoms and molecules theory", Vilnius, 1979. Thesis of the report, 4.2

12.	Al_2O_3-γ+C_2H_5OH.	IR and CNDO/2M methods. It was stated, that O...H bond is responsible for physical desorption, i.e. $ROH \rightarrow RO^- + H^+$ - its formal mechanism.	H^+	Korsunov V.A., Chuvyilkin N.D., Jidomirov G.N., Kazansky V.V., "Kinetics Catalysis", 1981, v. 22, p. 930.
13.	BF_3+EtOH	It was obtained, that the rate of polymerization of monomer or the rate of initiation of $BF_3 \cdot EtOH$ is approximately 10 times higher than for $BF_3 \cdot Et_2O$.	H^+	Polymerization of Styrene Catalyzed by a boron Trifluoride-Ethanol, Chem., 1963, v. 63, pp. 243-244. Jamoto M., Aoki S., Cationic Polymerization Catalyzed by boron Trifluoride-Ether complexes. Macromol. Chem., 1961, v. 48, pp. 72-79.
14.	BF_3+CH_3OH	It was studied equilibrium isomerization of buten-1 into cys- and trans-buten-2. Isomerization proceeds in parallel to protonation which includes the mechanism of formation of carbonium ions. This was shown with the help of deuterated acetic acid.	R^+	Roberts J.H., Katovic A.M. Eastham A.M. Catalysis of olefin isomerization of boron Trifluoride. J. Polymer Sci., A11, v. 8, pp. 3503-3510, 1970.
15.	H^++$R(CH$-$(OH)CH_3)$ $R=CH_3$, C_2H_5 i-C_3H_7, t-C_4H_9	Method CNDO/2. The effect of C_α-O bond weakening in protonated forms is stronger in comparison with pure secondary alcohols.	R^+	Sedlacer J., Kraus M. Effect of structure on catalytic dehydration of alcohols as modelled by quantum-chemical calculation. "Collect. Czech. Chem. Com.", 1976, v. 41, No. 1, pp. 248-250.

16.	$H^+ + ROH$ ($R = CH_3$, C_2H_5, i-C_3H_7, t-C_4H_9) $C^+H_3 + ROH$ $(CH_3)_3C^+ + ROH$	CNDO/2 method. C_α-O bond is weakened with the growth of alcohol radical branching.	R^+	Sangalov Ju.A., Minsker K.S., Ponomarev O.A., Babkin V.A., Quantum-chemical calculation of alcohols fragmentation in complexes with different acids.- TECh, 1983, v. 19, No. 5, pp. 615-620.
17.	$AlCl_3 + CH_3OH$ (C_2H_5OH, i-C_3H_7OH t-C_4H_9OH, C_6H_5OH)	At T=-78°C the following complexes are formed: $4ROH \cdot AlCl_3$, $2ROH \cdot AlCl_3$, $ROH \cdot AlCl_3$ and $ROH \cdot Al_2Cl_3$. Complexes are equally ionized, and their ability to ionize polymeization of isobutylene is proportional to alcohols acidity, which participate in the complex composition. Phenol behave itself in another way: the of complexes with phenol>>than that of complexes with aliphatic alcohols. The break at OPh is higher than at OR. The following reactions are possible: 1) 1:2 $ROH + 2AlCl_3 \rightleftarrows H^+[Al_2Cl_6 \cdot OR]^-$ 2) 1:1 $ROH + AlCl_3 \rightleftarrows H^+[AlCl_2]^+ \cdot [AlCl_3 \cdot OR]^-$ 3) 2:1 $ROH + H^+[AlCl_3 \cdot OR]^- \rightleftarrows [ROH_2]^+[AlCl_3 \cdot OR]^-$ 4) 4:1 $2ROH + [ROH_2]^+[AlCl_3 \cdot OR]^- \rightleftarrows [2ROH \cdot ROH_2]^+[AlCl_3 \cdot OR]^-$	H^+	Zlamal Z., Kazda A., Donor-acceptor Interactions in Cationic Polymerization. V Influence of some alcohols on the molecular weight of polyisobutylene in polymerization catalyzed by aluminium trichloride. J.Pol. Sci., A(1), v. 10, pp. 3199-3207, 1963.

18.	$AlCl_3+EtOH$	It is shown the connection of electric conductivity of $AlCl_3 \cdot EtOH$ solution and the molecular mass of polyisobutylene.	H^+	Zlamal Z., Kazda A. The role of nonpolar compounds in cationic polymerization of isobutylene. J. Polym. Sci., 191, v. 53, pp. 203-208
19.	$AlCl_3+CH_3OH$ (C_2H_5OH, C_4H_9OH)	It was shown that electric conductivity of $AlCl_3$ solutions increases in the sequence: $CH_3OH<C_2H_5OH<C_4H_9OH$.	H^+, R^+	Golub A.M., Kan Cha Pham, Samoilenko Y.M. Electrolythic properties of chloride-aluminium. J.Phys. Chem., 1970, v. 44, No. 11, p. 2779.
20.	BF_3+CH_3OH	It was studied the properties of isobutylene oligomers (polymerization degree≈10). $BF_3 \cdot ROH+(CH_3)_2C=CH_2 \rightarrow BF_3 \cdot OR^- +(CH_3)_3C^+$ - initiation. $$\sim(CH_2-\overset{\overset{CH_3}{\textstyle\vert}}{\underset{\underset{CH_3}{\textstyle\vert}}{C}})_{\gamma+1}+BF_3\cdot OR^- \rightarrow BF_3 \cdot ROH+$$ $$+\sim(CH_2-\overset{\overset{CH_3}{\textstyle\vert}}{\underset{\underset{CH_3}{\textstyle\vert}}{C}})_\gamma-CH_2-\overset{\overset{CH_2}{\textstyle\vert}}{C}-CH_3-break.$$	H^+	Dainton F.S., Sutherland G.B. Infrared analysis in elucidation of the BF3-catalyzed vapour-phase polymerziation of isobutylene. J.Pol.Sci., 1949, v. 4, pp. 37-43.

21.	BF_3+CH_3OH (C_2H_5OH, n-C_3H_7OH, i-C_3H_7OH).	Alkylation of phenol was performed at T=170°C, CH_3OH, C_2H_5OH. Anizol phenol was obtained. In the case of C_3H_7OH and i-C_3H_7OH the mixture of isopropylphenyl ester with 2,4-diisopropylphenyl and isopropyl ester $BF_3 \cdot 2ROH \rightarrow H^+[HO \cdot BF_3]^- +ROR$ (at 100-138°C) $ROR+H^+[HO \cdot BF_3]^- \rightleftharpoons R\text{-}O\text{-}R+H_2O$ $R_2O \cdot BF_3 \rightarrow R^+ +[RO \cdot BF_3]^-$ $BF_3+ROH \rightarrow BF_3 \rightleftharpoons H^+[RO \cdot BF_3)^-$	H^+	Romadan I.A. Alkylation of phenol by complexes of alcohols with boron fluoride. Catalytic transformations of hydrocarbons. Col. Sci. Pap. 1, Jdanov's Irkutsk GU, 1980, pp. 109-115.
22.	BF_3+CH_3OH	Investigation of kinetics of butenes at -20÷+20°C points out the protonation with the formation of classic carbonium ion	H^+	Roberts J.M., Katovic Z., Eastham A.M., Catalysis of olefins isomerization by boron trifluoride.
23.	$BF_3 \cdot 2CH_3\text{-}COOH$ (adamantol)	For adamantols the main role in transformation of complexes is played by the formation of adamantylcations, being stabilized by means of secondary reactions.	R^+	Shokova E.A., Garicheva O.N. Adamantols 1-2 trasnformation in presence of CH3COOH, CF3COOH and BF3◊2CH3COOH Neftekhimia, 1975, v 15, No. 3, pp. 363-368.

| 24. | $SnCl_4 + \ldots$ | It forms G-complexes by 0-0 of phenol, which are precisely these the initiators of polymerization | H_+ | Bauer R.F., LaFlair R.T. Russel K.E. Complexes of Stannic Chloride and Alkyl Phenols and Influence of these complexes and triphenols on the cationic polymerziation of isobutylene. Can. J. Chem., v. 48, pp. 1251-1262, 1970. Bar R.F., Russel K.E. Cationic polymerization of isobutylene initiated of stannic chloride and phenols. Polyer end group studies. J. Pol. Sci., 1971, A1(9), pp. 1451-1458. |

CHAPTER 5

APPLIED QUANTUM CHEMISTRY OF CATIONIC POLYMERIZATION OF OLEFINS

5.1. Indexes of reactability in cationic polymerization of olefins.

The first stage of the applied quantum chemistry is the special approach, basing on the idea about indexes of reactibility of the molecules [55], and in particular, of complex Lewis and Brenstedt acids, which play the role of catalysts of cationic polymerization. As the reactibility of any molecule it is meant its ability to participate in one or another reaction. That is why each time it is required a note, what special reaction is meant. As the rate constant is the accurate measure of the reactibility, i.e. the value, which is the function of features of several molecules, and not a single one, for the reaction of order higher than unit, then their reactibility is the value concretely semiquantitative, characterizing the ability of the molecule to participate in a number of uniformal reactions, in particular, in reaction of electro- and nucleophilic substitution, radical attachment, etc.

Consequently with semiquantitative character of the notion of the reactibility of a molecule it is impossible to give fully correct definition of an item, which is assumed to call its index [55]. Nevertheless, it is necessary to point out, that the reactibility index is any feature of electron structure of a molecule, for which correlation with the reactibility of a group of chemical compounds in the concrete class of reactions is set [55].

The simplest index of reactibility of the molecules is the charge on atom q_A. It is reliably set the correlation between its positive values with the rate of reactions of nucleophilic as well as negative ones – with electrophilic attachment or substitution [126].

Experimental tests of numerical values of q_A is not easy. Coincidence of the calculated dipole moment of a molecule with that measured experimentally is just an indirect and meant proof of the probability of the charge distribution. The method of photoelectronic microscopy [127][128] allows to estimate this distribution in more detail.

Particularly, the following quantum-chemical parameters were tested for the possibility to be the index of reactibility of complex catalysts of cationic polymerization of $R_nAlCl_{3-n} \cdot H_2O$, (HCl, ROH) $R_nBF_{3-n} \cdot H_2O$ (HCl, ROH) $R_nMgCl_{2-n} \cdot H_2O$ (ROH): E_{tot}-total energy of the system; E_{bond} - total energy of bonds; $\Delta E_{det\,ach}^{H^+}$ - energy of proton detachment from proton-donor of the complex; E_g - energy gain (or loss) resulting the reaction; q_A (q_{Al}, q_B, q_{Mg}, q_F, q_{Cl}, q_O, q_C, q_H, etc.) - charges on atoms of the complex studied; P_{AB} - bond order; D - dipole moment; E_{uoMO}, E_{lfMO} - energy of upper occupied and lower free Molecular Orbitals (the method of border orbitals; Δq - a part of charge translated from donor to acceptor.

The calculation of nearly 300 objects of the models of complex Lewis and Brenstedt acids (Tables 4.2, 4.10, 4.17, 4.20, 4.29, 4.30), the analysis of theoretical and experimental works allowed to obtain the validity of correlation dependence of pKa –

the universal index of acidity, on q_H – the charge on hydrogen atom for the initiators of electrophilic polymerization of olefins, suggested above in [4] for more simple H-acids. However, this correlation dependence does not work in the case of strongly acidic complex catalysts, in which proton-donors participate, containing d-orbitals, $R_nAlCl_{3-n} \cdot HCl$ for example. Here the informative factor was found in the energy of $\Delta E_{detach}^{H^+}$ proton detachment from the proton-donor of HCl in the complexes with aluminum chlorides. That is why it was used the correlation, obtained for carbonium acids: $\delta pKa \equiv \beta \delta E_{detach}^{H^+}$, which is valid for $R_nAlCl_{3-n} \cdot HCl$ complexes, as it was shown in [109]. Sufficiently reliable correlation with the experimental data about the activity of the complexes of alcohols with Lewis and Brenstedt acids is given by bonds orders P_{AB}.

Thus, it can be stated that the following quantum-chemical parameters are playing the role of the indexes of reactibility at cationic polymerization of olefins: q_{H^+} – the charge on hydrogen atom of the proton-donor of complex catalysts (for weakly acidic initiators), P_{AB} – bond orders and $\Delta E_{detach}^{H^+}$ – the energy of detachment of a proton H^+ from the proton-donor in the complexes with Lewis acids (R_nAlCl_{3-n}, R_nMgCl_{2-n}, R_nBF_{3-n}).

5.2. Cationic polymerization of olefins as conjugated process.

The performed quantum-chemical calculation of the models of the complex acids of Lewis (aluminum chlorides, magnesium chlorides, boron fluorides) and Brenstedt (hydrogen chloride, hydrogen fluoride, water, alcohols) points out the dependence of acidic properties on chemical structure of proton-donor. The complexes of hydrogen chloride (hydrogen fluoride) with Lewis acids are inclined to ionization. In this case the initiation of cationic polymerization of olefins is connected with the particular interaction between an olefin and a base for the proton attachment. The interaction of strong complex acids – $R_nAlCl_{3-n} \cdot HCl$ ($R_nBF_{3-n} \cdot HF$) – with a base (an olefin) is of nonbarrier character. Here an important role is devoted to orbital interactions. Complexes of weak proton-donors (H_2O, ROH) are not AC of cationic polymerization. Calculation of the complexes of aluminum chlorides (magnesium chlorides, boron fluorides) with water and alcohol points out the important role of the olefin coordination at the initiation of cationic processes under the influence of weak complex catalysts in halogenic nonpolar media. Similar to heterogeneous systems [93], in the case of which the olefin coordination supposes the orientational influence of catalytic surface, in homogeneous systems it includes active roles in the performance of acidic properties of the catalyst.

On the initial stages of the interaction of monomer with the complexes $R_nAlCl_{3-n}(R_nMgCl_{2-n}$, $R_nBF_{3-n}) \cdot H_2O$ (ROH) it is observed their mutual orientation, characterized by a definite angle of olefin attack by a proton from the complex and by the change of dislocation of active bond $O\text{-}H_{(1)}$ of the proton-donor according to the olefin during the reaction. It is sufficient, that during the process of interaction of the monomer with the complexes aluminum chlorides (magnesium chlorides, boron fluorides) – proton-donor (water, alcohols) the sufficient increase of Brenstedt's acidic strength of the complex takes place (δpKa increases from +7 to -15). As a consequence of this the complex acquires the properties of strong H-acid. However, proton transfer is not the limiting stage and is conjugated in time with the formation of counterion, which is promoted by preliminarily coordination of water (alcohol) with Lewis acid. It may be stated that the formation of counterion in its turn supports the proton transfer.

Thus, in the case of homogeneous weakly acidic catalysts the coordinational conjugated mechanism of initiation of electrophilic processes is probable, which has been pointed out before for heterogeneous systems [93].

The mechanism suggested, which is the logic continuation of the known sequence of cationic transformations of olefins, is shown on the Figure 5.1 (conjugated processes).

As well as the process of initiation of electrophilic polymerization, the reaction of contact ionic pair with olefin, i.e. the stage of the chain growth, is a conjugated process of interactions of the components of systems. At the attack of AC by a monomer counterion makes the approach to carbcationic centre difficult, providing the directivity of the further attack and mutual orientation of the components of the system. Anionic fragment of the contact ionic pair participates directly in electron transitions, connected with the disclosure of double bond of olefin. The character of the charges change on carbonic atoms of olefin, carbcation and oxygen atom of anion testifies the conjugated mechanism and oxygen atom of anion testifies the conjugated mechanism of the reaction. The weakening of carbcation-counterion bond is not accompanied by the expected increase of charge on the components of ionic pair and the increase of total energy of the system. Abversable separation of charges is compensated by the coordination of olefin in relation to active $C_{(1)}^{+} - O^{-}$ bond. The process is accompanied by a definite spatial dislocation of counterion, creating minimum obstacles at the AC attack by monomer.

1. Polymerization of propylene [2]:

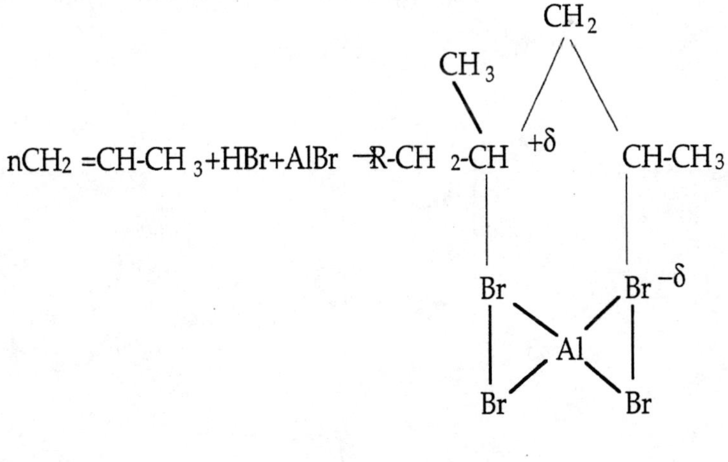

$$nCH_2 = CH\text{-}CH_3 + HBr + AlBr \rightarrow R\text{-}CH_2\text{-}CH \cdots CH\text{-}CH_3$$

Figure 5.1. Cationic transformations of olefins (continued on next page).

2. Isomerization of butene-1 into butene-2:

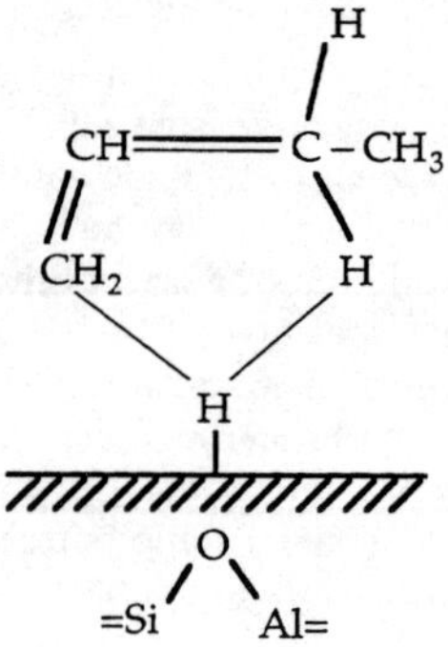

$$
\begin{array}{c}
\text{CH}=\!=\!\text{C}-\text{CH}_3 \\
\end{array}
$$

3. The break of cationic polymerization:

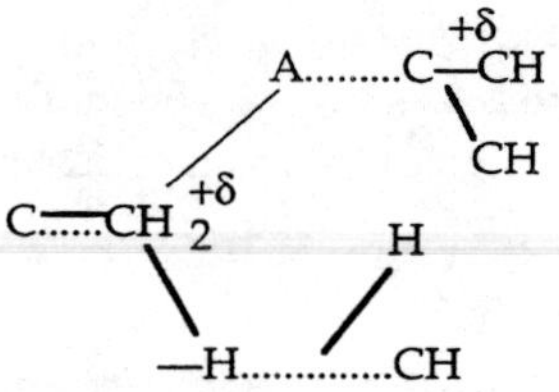

4. The suggested conjugated (cooperative, concert) mechanism of initiation of cationic polymerization of olefins (from the analysis of quantum-chemical calculations):

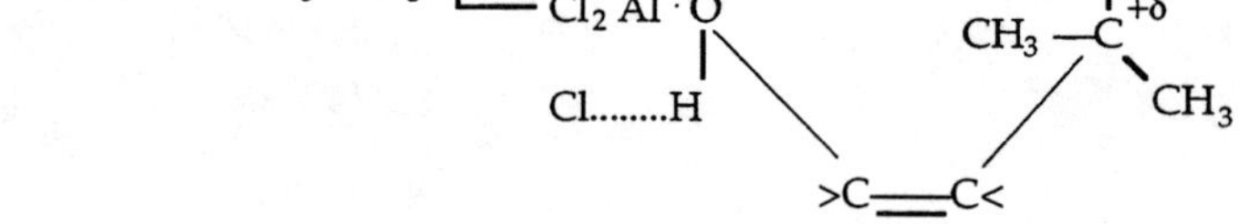

Figure 5.1. Cationic transformations of olefins.

As a result, the reaction of interaction of active centre with olefin is accompanied by total energy gain of 500-560 kJ/mole.

The suggested mechanism of initiation of cationic processes by the complex catalysts of metal halogenide – proton-donor and its dependence on the nature of the system may be used for the estimation of acidic strength of newly synthesized complexes in order to clear up the possibilities of their possessing catalytic properties. The parameters of acidity, obtained from the calculation of models of these complexes, are presented as the values of the uniformal acidity index through the known correlation expression ($q_H \equiv dpKa$).

The investigated complex will be the catalyst *apirori* at high δpKa values. In the case of low values of δpKa the final estimation of acidic-catalytic properties is performed experimentally or by calculations with regard to the substrate nature. The analysis of the role of base in its turn open the way to the explanation of selective properties of catalysts with low acidic strength. Consequently, the comparison of calculated and experimental data allows to analyze the connection of acidic and catalytic properties of the complexes metal halogenide – proton-donor.

5.3. The interconnection of acidic-catalytic properties of complex Lewis and Brenstedt acids.

The analysis of the totality of quantum-chemical calculations and experimental data about acidic strength and catalytic activity in electrophilic processes of polymerization of sufficiently stable complexes of halogenides, alkyls and alkylhalogenides of nontransitional metals (Al, B, Mg) with proton-donors (H_2O, ROH, HCl, HF) allows to make a conclusion, that acidic strength is the measure of activity and selectivity of the action of catalysts of this type. There were cleared up new important facts, promoting the deeper understanding of the role of acidic strength and the interconnection of acidic-catalytic properties of complex acids of Lewis and Brenstedt.

1. Complexes $R_nAlCl_{3-n} \cdot HCl$, $R_nBF_{3-n} \cdot HF$ are characterized by high acidic strength, independent on ligand surrounding of heteroatoms (Al, B, Mg) (pKa=-16÷-60) The detachment of a proton is energetically profitable. The complexes are inclined to spontaneous ionization and initiation of electrophilic polymerization.

2. Complexes $R_nAlCl_{3-n} \cdot H_2O$, $R_nMgCl_{2-n} \cdot H_2O$, $R_nBF_{3-n} \cdot H_2O$ and also the products of their transformations $OHAlCl_2 \cdot H_2O$, $AlCl_4O \cdot H_2O$, $R_nAlCl_{3-n} \cdot H_2O \cdot HCl$, $B_2F_4O \cdot H_2O$ are characterized by low acidic strength (pKa=+1.3÷+14.4). The proton detachment is energetically nonprofitable. The complexes of this type are not inclined to spontaneous ionization and are able to initiate electrophilic polymerization only in presence of bases (monomer, arene, etc.).

3. Similar to the corresponding aquacomplexes, R_nAlCl_{3-n} (R_nMgCl_2, R_nBF_{3-n})$\cdot$ROH complexes are characterized by low acidic strength (pKa=+1÷+15.1). Proton detachment is energetically nonprofitable. The complexes are not inclined to the spontaneous ionization. Complex-linked alcohols are able to perform proton-donor or carbcationic activity, depending on chemical structure of R radical in alcohol molecule, acidic strength of the complex forming agent, and at its low value – on specific interactions of reagents, unusual form of proton-donors coordination (H_2O, ROH) according to Lewis acid (specially concerning Mg compositions). This sufficiently influences the elementary stage of initiation of the process of electrophilic polymerization of olefins, and consequently, on the character of the end groups in the

forming macromolecules. In this connection is can be observed the dependence: the selectivity of the polymerization process increases with the increase of size of ligand of Lewis acid [131] or of R radical volume in the alcohol molecules (Table 4.26), participating in the complexes of halogenides of nontransitional metals with proton-donors.

Complexes of halogenides of nontransitional metals with weak proton-donors (H_2O, ROH) are united by their low acidic strength, which makes the fact of their catalytic activity performance without cocatalyst action of the third component (electrodonor) doubtful in the electrophilic processes, and its weak dependence on chemical nature of the initial Lewis acid. In this case the role of monomer as the base and its direct participation in the formation of active centres is determinant.

The method of estimation of acidic strength of complex catalysts R_nAlCl_{3-n} – proton-donor, suggested in [4], may be spread to the wider class of complexes, based on nontransitional metals. The analysis of pKa values of more than 300 compounds and complexes, which are potentially able to be catalysts of electrophilic processes of polymerization of olefins in a wide range of $+50 \leq pKa \leq -60$, allows to separate conditional zones of their activity and selectivity (Figure 5.2):

I. $+50 \leq pKa \leq +12$ sufficiently low values of pKa. According to this cause H-acids are inactive as catalysts of electrophilic processes of polymerization. They are, in particular, individual ROH and H_2O, $R_3Al \cdot H_2O$, $R_3Al \cdot ROH$, etc. (Tables 4.10, 4.17, 4.25, Figure 4.2).

II. $+12 \leq pKa+1$ – optimum pKa values for complex catalysts of electrophilic polymerization of olefins $C_2H_5AlCl_2 \cdot H_2O$, $Cl_3Al \cdot H_2O$, $Cl_2Mg \cdot H_2O$, $C_2H_5AlCl_2 \cdot ROH$, etc. (Tables 4.10, 4.25, 4.17). The catalysts possess high activity and selectivity.

III. $+1 \leq pKa \leq -60$ – High value of pKa, at which H-complexes are sufficiently active, but low selective: $R_nAlCl_{3-n} \cdot HCl$, $R_nBF_{3-n} \cdot HF$, protonated alcohols, etc. (Tables 4.2, 4.6). It should be mentioned, that in general case border values of pKa may transfer to one or another side along pKa axis (Figure 5.2), depending on the compounds class and the list of other causes.

The analysis of the results of quantum-chemical calculations and the data of experimental investigations allows to formulate the following rule:

The increase of acidic strength of catalysts of electrophilic reactions of polymerization promotes the growth of their activity, but decreases selectivity of the process.

Conclusion.

Thus, there is a definite progress in the sphere of fundamental investigations in the chemistry of polymers. The knowledge about the mechanism f initiation and growth of cationic polymerization of olefins were deepened as well as about the structure of active centres, active role of counterion. The important role in understanding these mechanisms is devoted to quantum-chemical calculations, which predictive force allows to solve serious practical tasks, for example, the selection of high temperature catalysts for some cationic processes. In this case salty complexes $M^{+\delta}$ $[RAlCl_3]^{-\delta}$, similar to stable $M^{+\delta}[AlCl_4]^{-\delta}$ ones by their nature, are interesting, which decompose with MCl elimination [129]. Nevertheless, the calculation pointed out that complex decomposition by AC-C bond of nonsymmetric ion is preferable, that is characteristic also for other counterions with different named ligands (R_2AlCl_2). On the other hand, it was found by the calculation, that the interaction of complex cation with external acceptor can efficiently compete with this reaction. This interaction becomes more preferable at the decrease of the cation size and taking into account solvatation factor. The experimental data, pointing out the possibility of substitution of alkali cation in the complexes by another ones (H^+, C^+) with the initiation of cationic processes, are in connection with these results. In this case direction of the process, caused by the complexes, is sufficiently influenced by substrate nature. In more rigid conditions there are also possible other reactions of the external-spheric cation of salty complexes, for example with simple carbon-carbonic bonds. The above mentioned played the role of the basis for experimental testing of cationic properties of the complexes during polymers degradation. These experiments were completed by the development of efficient catalyst of cationic depolymerization of polyisobutylene [130].

The results of quantum-chemical calculations are also applied for the solution of other tasks, being important from the practical point of view:

1. Estimation of acidic strength of Lewis complexes (R_nAlCl_{3-n}, R_nMgCl_{2-n}, R_nBF_{3-n}) and Brenstedt ones (H_2O, ROH, HCl, HF) in order to clear up the probability for them to perform catalytic properties.

2. Systematization and explanation of literature data about the behavior of complexes of alcohols with different acids in cationic processes, being interesting for obtaining polyolefins with the required end groups.

3. Preliminary estimation of selectivity of the complex catalysts of cationic polymerization of olefins (Figure 5.2).

4. Experimental selection of new complex catalysts of electrophilic processes, in particular selective depolymerization of polyisobutylene and BC.

Moreover, the general analysis of quantum-chemical calculations of Lewis and Brenstedt complexes and the experimental data allowed to make a conclusion: the increase of acidic strength of catalysts of electrophilic reactions of polymerization promotes the increase of their activity, but decrease the selectivity of the process.

All this testifies about the fact, that semiempyric quantum-chemical methods CNDO/2 and MINDO/3, as the most admissible ones despite their shortcomings and approximations for the calculations of models of complex catalysts of cationic polymerization, began to perform their main application – searching for new regularities, dependences, correlations of quantum-chemical parameters of practically important complexes with their experimental data.

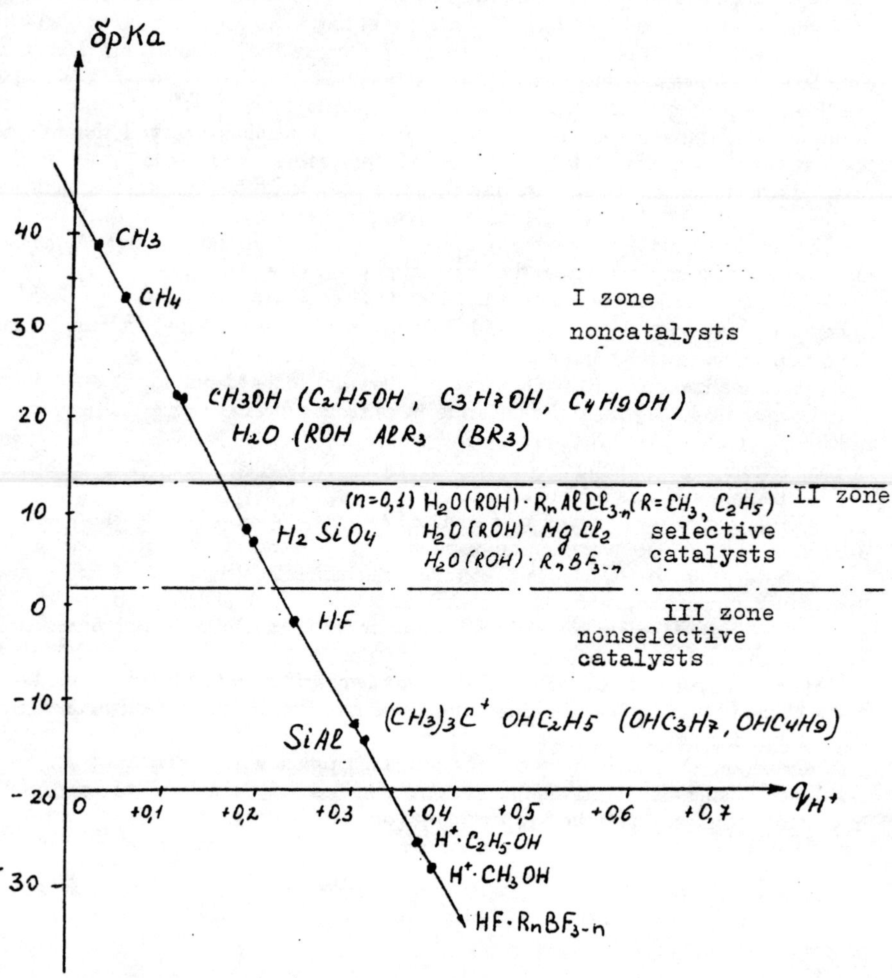

Figure 5.2. The dependence of charge on the hydrogen atom f proton-donor on universal index of acidity (-.-.- – conditional line of selectivity zones) (CNDO/2 method).

REFERENCES

1. H.A. Burson, H. Staudinger, **Ind. End. Chem.** 18, 381 (1966).
2. J. Kennedy, **Cationic Polymerization of Olefins**, Moscow, Mir (1978).
3. G.A. Olah, **Fridel-Crafts Chemistry**, N.Y., J. Wiley and Sons, 1973.
4. W.Grabowski, M. Misono, Yu. Yoneda, **J. Catal.** 61, 103 (1980).
5. S. Enikolopyan, E.F. Oleinik, **Encyclopedy of Polymers**, 1, 973 (1972) (in Russian).
6. G. Oudian, **Polymer Chemistry**, Moscow, Mir, 1974.
7. Reactivity, Reaction **Mechanisms and Structure in Polymer Chemistry**, Ed. A. Jenkins, A. Ledvis, Moscow, Mir, 1977.
8. **Cationic Polymerization**, Ed. P. Plesh, Moscow, Mir, 1966.
9. Yu.A. Sangalov, V.A. Babkin, **Acidity and Catalytical Properties of Metal Protondonor Complexes**, Ufa, Bashkirian University, 1989.
10. B.L. Erusalimskii, S.G. Lyubetskii, **Processes of Ionic Polymerization**, Leningrad, Khimiya, 1974 (in Russian).
11. B.L. Erusalimskii, **Ionic Polymerization of Polaric Monomers**, Leningrad, Nauka, 1969.
12. **Second International Symposium on Cationic Polymerization.** October 1973, Rouen, France/Macromol. Chem., 1974, 175, 1017 (1974).
13. Fourth International Symposium on Cationic Polymerization. June 20-24, 1974, Akron, Ohio/J. **Polym. Sci.**, Symposium, 56, 1 (1976).
14. K. Tanabe, **Solid Acids and Alkalies**, Moscow, Mir, 1973.
15. J.P. Kennedy, E. Melby, J. Johnston, **J. Macromol. Sci.** (chem), A8, 463 (1974).
16. **Modern Aspects of Chemistry of Carbonic Ions**, Ed. V.A. Kaptyug, Novosibirsk, Nauka, 1975.
17. Petrova et al, **Doklady Akademii nauk SSSR**, 233, 602 (1977).
18. I.M. Alpatova, V.V. Gavrilenko, Yu.M. Kessler, O.P. Osipov, D.N. Maslin, **Complexes of Metalorganic Hydrid and Halloid Aluminum Compounds**, Moscow, Nauka, 1970.
19. Yu.A. Sangalov et al, **Teoreticheskaya i eksperimentalnaya khimiya**, 13, 623 (1977).
20. T. Higashimura, T. Watanabe, S. Okamura, **Chm. High Polym.** Japan, 20, 680 (1963).
21. M.F. Shostakovskii et al. **Izvestia Akademii nauk SSSR, ser. khimicheskaya**, 10, 1848 (1964).
22. S. Aori, H. Nakamura, T. Otsu, **Macromol. Chem.** 115, 282 (1968).
23. Yu.B. Yasman et al. **Zh. organicheskoi khimii**, 26, 1665 (1976).
24. G.A. Razuvaev et al., **Doklady Akaemii nauk SSSR**, 239, 338 (1978).
25. Yu.A. Sangalov et al. ibid, A20, 1331 (1978).
26. Yu.A. Sangalov et al. ibid, A20, 1331 (1978).
27. Yu.B. Yasman et al. ibid, B21, 5667 (1979).
28. G.G. Prekampus, **New Problems of Physical Organic Chemistry**, Moscow, Mir, 1969, p. 257-334.
29. Yu.A. Sangalov et al. **Doklady Akademii nauk SSSR**, 265, 671 (1982).
30. A.A. Rabinovich, Z.Ya. Khavin, **Short Chemical Handbook**, Leningrad, Khimiya, 1977, p. 36, 40 (in Russian).
31. J. Kennedy, A.V. Langer, **Uspekhi Khimii**, 36, 77 (1967).
32. R.O. Colclough, F.S. Dainton, **Trans. Faraday Soc.**, 54, 886 (1958).
33. E.B. Lyudvig et al. **Doklady Akademii nauk SSSR**, 1566, 1163 (1964).

34. T.B. Bogomolova, A.P. Gantmakher, ibid, 217, 369 (1974).

35. T.B. Bogomolova, A.P. Gantmakher, ibid, 230, 117 (1976).

36. T.B. Bogomolova, A.P. Gantmakher, **Vysokomolekulyarnue soedinenia**, 20A, 1315 (1978).

37. T. Aoyagi, T. Araki, H. Tani, **Makromol. Chem.** 163, 45 (1973).

38. V.D. Petrova et al. **Izvestia Akademii nauk SSSR, ser. khimicheskaya**, 6, 1373 (1978).

39. Yu.A. Sangalov et al. **Vysokomolekulyarnye soedineniya**, 21A, 2267 (1979).

40. I.I. Ermakova et al. ibi, A11, 1639 (1969).

41. Yu.A. Sangalov et al, ibid, A20, 1331 (1978).

42. F.S. Dainton, B.B.M. Sutherland, **J.Pol. Sci.**, 4, 37 (1949).

43. I.A. Romadin, **Alkylation of Phenol by Alcohol-Fluorine Borine Complexes**, Irkutsk, Irkutsk State University Publ., 1980, p. 109-115 (in Russian).

44. J. Derouault, T. Dziembowska, M.T. Florel, **J.Mol. Struc.**, 47, 59 (1978).

45. S. Aori, M. Imoto, **Makromol. Chem.** 65, 243 (1963).

46. S. Aori, M. Imoto, ibid, 48, 932 (1961).

47. H.R. Allock, A.M. Eastham, **J. Chem. Soc.**, 13, 30 (1966).

48. T. Szell, A.M. Eastham, **J. Chem. Soc.**, 13, 30 (1966).

49. J.M. Roberts, L. Katolic, A.M. Eashem, **J. Polym. Sci.**, A1, 3503 (1970).

50. Z. Zlamal, L. Ambroz, K. Vesely, ibid, 24, 285 (1957).

51. Z. Zlamal, A. Kazda, ibid, A1, 3199 (1963).

52. E.A. Shokova et al., **Doklady Akademii nauk SSSR**, 212, 386 (1973).

53. Catalytic Properties of Substances, Ed. V.A. Roiter, Kiev, **Naukova Dumka**, 1968.

54. A. Gandini, H. Cheraclome, **Advances in Polym. Sci.**, 34/35, 1-279 (1980).

55. Yu.E. Eizner, B.L. Erusalimskii, **Electronic Aspect of Reactions of Polymerization**, Leningrad, Nauka, 1976.

56. P.H. Plesh, **Polymrytworz. Wielkoczasteczk.** 18, 557 (1973).

57. N. Smid, **Uspekhi khimii**, 17, 799 (1973).

58. I.P. Beletskaya, A.A. Solov'yanov, Zh. **Vsesoyuznogo khimicheskogo obshchestva**, 22, 286 (1977).

59. M.V. Bazilevskii, **Method of Molecular Orbits and Reactivity of Organic Molecules**, Moscow, Khimiya, 1969.

60. V.A. Gubanov, V.P. Zhukov, O.A. Litinskii, **Semiempyric Methods of Molecular Orbitals in Quantum Chemistry**, Moscow, Nauka, 1969.

61. D. Heartry, **Calculations of Atomic Structures**, Moscow-Leningrad, Mir, 1960 (in Russian).

62. L.A. Blyumenfeld, A.K. Kukushkin, **Course of Quantum Chemistry and Structure of Molecules**, Moscow, Moscow State University Publ., 1980.

63. L. Culike, **Quantum Chemistry**, Moscow, Mir, 1976.

64. V.I. Minkin, B.Ya. Simkin, R.M. Minyaev, **The Theory of Structure of Molecules**, Moscow, Vysshaya shkola, 1979.

65. M.J. Duar, **The Theory of Molecular Orbitals in Organic Chemistry**, Moscow, Mir, 1972.

66. T.M. Zhidomirov, A.A. Bagatur'yants, I.A. Abronin, **Applied Quantum Chemistry**, Moscow, Khimiya, 1979.

67. I.B. Bersuker, **Structure and Properties of Coordinative Compounds**, Moscow, Nauka, 1974.

68. J. Seegal, **Semiempyric Methods of Electronic Structure Calculations**, Moscow, Mir, 1980.

69. M.E. Dyatkina, **Fundamental Theory of Molecular Orbitals**, Moscow, Nauka, 1975.

70. O.K. Davtyan, **Quantum Chemistry**, Moscow, Vysshaya shkola, 1978.

71. J. Slater, **Electronic Structure of Macromolecules**, Moscow, Mir, 1965.

72. D.R. Armstrong, P.G. Perkins, J.P. Stewart, **J. Chem. Soc. Dalton. Trans.** 8, 103 (1973).

73. G.E. Zaikov, **J. of Pol. Mater.**, 24, 1 (1994).

74. P.V. Schastnev, G.A. Bogdanchikov, **Teoreticheskaya i eksperimental'naya khimiya**, 16, 667 (1975).

75. V.H. Kondrat'ev, E.E. Nikitin, A.I. Reznikov, S.Ya. Usmanskii, **Thermal Biomoleculare Reactions in Gases**, Moscow, Nauka, 1976.

76. M. Harigatai, I. Harigatai, **Geometry of Coordinative Compounds Molecules in Vapor Phase**, Moscow, Mir, 1976.

77. M.V. Basilevskii, Zh. **Vsesoyuznogo khimicheskogo obshchestva**, 22, 245 (1977).

78. O. Navaro, E. Blaiten-Barojos, E. Clement et al., **J. Chem. Phys.** 65, 2337 (1978).

79. K.B. Wiberg, **Tetrahedron**, 24, 1083 (1968).

80. I.P. Zakharov, A.O. Litinskii, Zh. **Strukturnoi khimii** 24, 111 (1983).

81. I.P. Zakharov, A.O. Litinskii, V.A. Tolstonogov, ibid., 25, 3 (1984).

82. E. Silla, E. Scrocco, J. Tomasi, **Theor. chim. acta.** 40, 343 (1975).

83. P. Hallpap, G. Heublein, **Z. chem.**, 17, 446 (1977).

84. G. Alagona, E. Scracco, E. Silla, **Theor. Chem. acta**, 45, 127 (1977).

85. H. Umeayma, **Chem and Pharm. Bull**, 28, 1633 (1980).

86. V.N. Plakhotnik et al., **Teoreticheskaya i eksperimental'naya khimiya**, 25, 123 (1989).

87. E.N. Gur'yanov, Gol'shtein, I.P. Romm, **Donor-acceptor Bond**, Moscow, Khimiya, 1973.

88. P. Hallpap, G. Heublein, V.Ja. Bogomoini, Yu.Ye. Eizner, B.L. Erusalimskii, S.S. Skorokhodov, **Europ. Polym. J.** 14, 1027 (1978).

89. H. Imai, T. Soegusa, J. Furkawa, **Makromol. Cem.** 81, 92 (1961).

90. O.X. Poleshchuk, Yu.K. Maksyutin, **Uspekhi Khimii**, 45, 2097 (1976).

91. K.Z. Gumargalieva, G.E. Zaikov, Yu.V. Moiseev, **Uspekhi khimii** (Russian Chemical Review), 63, 905 (1994).

92. N.D. Mikheikin, G.M. Zhidomirov, V.B. Kazanaskii, **All Union Conference on Mechanism of Catalytic Reactions, Abstracts of reports**, Moscow, vol. 2, p. 3 (1978).

93. V.B. Kazanskii, **Kinetika i kataliz**, 18, 1179 (1977).

94. Energy of Chemical Bonds Rupture, **Potential of Ionization and Electron Affinity** Ed. V.I. Kondrat'ev, Moscow, Academy of Sciences of the USSR Publ., 1962.

95. Yu.N. Gorlov, et al., **Teoreticheskaya i eksperimental'naya khimiya**, 24, 532 (1988).

96. M.Yu. Gorbachev, N.B. Bersuker, ibid, 24, 318 (1988).

97. N.D. Chuvylkin et al, **Kinetika i kataliz**, 14, 943 (1973).

98. N.D. Chuvylkin et al., ibid, 16, 92 (1976).

99. N.N. Sechenya et al., ibid, 21, 1184 (1980).

100. J. Sedlacer, M. Kraus, **Collect. Czech. Chem. Communs.**, 41, 248 (1976).

101. V.A. Korsunov et al., **Kinetikia i kataliz**, 22, 930 (1981).

102. N.A. Lygina et al, In book: **Catalytic Transformation of Carbohydrates**, Irkutsk, Irkutsk State University Publ., 1980, p. 20-24.

103. Yu.A. Sangalov et al., **Vysokomolekuylarnue soedineniya**, A22, 1588 (1980).

104. K.S. Minsker, Yu.A. Sangalov, **Izobutilene and its Polymers**, Moscow, Khimiya, 1986.

105. J.P. Kennedy, J.K. Gillham, **Adr. Polymer. Sci.**, 10, 1 (1972).

106. G.J. Sleddon, **Chem. and Ind.**, 37, 1492 (1961).

107. **Handbook on Chemistry**, Leningrad, Khimiya, vol. 4, p. 970 (1967).

108. Yu.A. Sangalov, et al, **Teoreticheskaya i eksperimental'naya khimiya**, 18, 535 (1982).

109. Yu.A. Sangalov, Yu.B. Yasman, Yu.Ya. Nelkenbaum, V.A. Babkin, **Acta Polymerica**, B38, 134 (1987).

110. K.S. Minsker et al., **Teoreticheskaya i eksperimental'naya khimiya**, 16, 181 (1980).

111. K.S. Minsker et al., **Plasticheskie massy**, 9, 31 (1988).

112. K.S. Minsker, V.A. Babkin, G.E. Zaikov, **Plast. Technolog. End.**, 29, 43 (1990).

113. S. Pasynkiewich, M. Baleslawski, A. Sadownik, **J. Organometal. Chem.**, 113, 303 (1976).

114. V.A. Babkin, **About Aqua-Complex Alumosiloxane**, Ufa, Bashkirian State University Publ., 1980..

115. J.P. Kennedy, A. Shinkawa, F. Williams, **J. Polym. Sci.**, A1, 1551 (1971).

116. P. Hallpap, G. Heublein, **Z. Chem.** 1, 21 (1976).

117. V.D. Petrova et al., **Zh. obshchei khimii**, 48, 1862 (1978).

118. A.A. Furman, **Nonorganic Chlorides**, Moscow, Khimiya, 1980..

119. C.P. Ivanova et al., **Vysokomolekulyarnye soedineniya** A28, 1217 (1986).

120. A. Gordon, R. Ford, **Companion of Chemist**, Moscow, Mir, 1975.

121. **Modern Aspects of Physical and Organic Chemistry**, Ed. E.M. Arnett, Moscow, Mir, 1967.

122. Yu.A. Sangalov et al., **Teoreticheskaya i prikladnaya khimiya**, 19, 615 (1983).

123. M. Attia, F. Cocace, G. Giranni, P. Giscomello, **J. Chem. Soc. Commun.** 21, 938 (1978).

124. R.F. Bauer, K.E. Russel, **J. Pol. Sci.**, 8, part A1, 1451 (1971).

125. V.V. Kharitonov, B.L. Psikha, G.E. Zaikov, **Intern. J. Polymeric. Mater.** 26, 121 (1994).

126. K. Hibasi, H. Baba, A. Rembaum, **Quantum Organic Chemistry**, Moscow, Mir, 1967.

127. K. Siegbahn, C. Nordling, G. Jachenson et al., **ESCA Applied to Free Molecules**, Amsterdam-London, North-Holland Publ. Co., 1969.

128. R.L. Braginskii, V.I. Nefedov, **X-ray Determination of Atoms Charge in Molecules**, Moscow, Nauka, 1966.

129. Yu.A. Sangalov, et al., **Teoreticheskaya i eksperimental'naya khimiya**, 15, 506 (1989).

130. K.S. Minsker, Ya. Sanagalov, Yu.A. Prochukhan, Al.Al. Berlin, P.P. Muslukhov, K.V. Prokof'ev, V.S. Kazanskii, C.R. Ivanova, V.A. Babkin, **Methods of Isobutylene Obtaining**, Certificate of the USSR, No. 330 80 13/04 of March 11, 1981.

131. K.S. Minsker, V.A. Babkin, G.E. Zaikov, **Polymer Yearbook**, 10, 107 (1994).

132. K.S. Minsker, V.A. Babkin, G.E. Zaikov, ibid, 8, 59 (1992).

133. K.S. Minsker, V.A. Babkin, G.E. Zaikov, **Polymer Plast. Technology and Engineering**, 29 (1-2), 43 (1990).

134. K.S. Minsker, V.A. Babkin, G.E. Zaikov, **Plastics**, 9, 1 (1988).

INDEX

—A—

acidity, xi, 32, 44, 47, 48, 54, 61, 62, 63, 65, 66, 69, 71, 75, 87, 92, 99, 103, 105, 113, 120, 122, 127, 129, 139, 150

Activation energy E_R, 5

Activity, 2, 3, 101, 106, 109, 114, 118, 121

Adamantols, 128

aliquot bond, 3

Alkaline Metals, 55

Aluminum Chlorides, 49, 61

aluminum halogenides, 64, 66

aprotic acids, 2

aquacomplexes, 33, 36, 37, 47, 61, 62, 63, 64, 65, 66, 67, 71, 72, 75, 83, 84, 86, 87, 90, 92, 93, 94, 107, 113, 117, 142

Arrhenius's activation energy, 37, 38

—B—

Basicity, 45, 97

benzene, 3, 5, 44, 45, 84, 122

Boron fluoride, 54, 116

—C—

carbcation, 5, 6, 7, 8, 52, 53, 74, 75, 76, 77, 78, 79, 81, 104, 107, 120, 136

Carbonium ion, 122

Carbonium salt, 2

catalytic activity, 49, 50, 62, 69, 92, 139, 143

cationic polymerization, xi, xii, 1, 2, 3, 4, 5, 6, 7, 8, 9, 23, 42, 44, 46, 49, 50, 51, 53, 55, 70, 75, 82, 83, 84, 93, 103, 107, 108, 111, 116, 119, 120, , 125, 127, 129, 132, 133, 138, 145, 148, 149

Chain growth, 4, 54

charge distribution, 24, 44, 126

cocatalyst, 2, 6, 7, 100, 113, 143

Complete Differential Overlapping (CDO), 22

Complete Neglecting of Differential Overlapping (CNDO), 22

Contact ionic pairs, 7

Coulomb energy, 44

Coulomb integrals, 40

counterion, 5, 7, 8, 51, 52, 53, 55, 58, 59, 60, 75, 76, 77, 79, 80, 81, 82, 83, 84, 90, 107, 108, 113, 114, 120, 121, 136, 145

cyanamydes, 2

cyclopropanes, 41, 42

cyclopropen, 2

—D—

dehydration, 49, 123, 125, 126

Dewar method, 45

—E—

electrical conductivity, 5, 44, 59

electron density, 3, 6, 45, 46, 53, 75

electron energy, 44

electrophilic reactions, 5, 50, 67, 71, 145, 149

electrostatic potential, 44

ethylene, 5, 51, 71, 76, 77, 78, 79, 80, 81, 82, 83, 108, 111, 113, 119

—G—

Gustavsson complexes (GC), 59

—H—

halogens, 2

Huckel method, 49

Hydrogen Chloride, 49

Hydrogen Fluoride, 54

—I—

isobutylene, 2, 39, 51, 52, 53, 54, 55, 56, 57, 58, 60, 61, 62, 66, 67, 70, 71, 72, 73, 74, 75, 81, 82, 83, 84, 87, 89, 90, 91, 93, 104, 105, 106, 107, 108, 111, 113, 119, 120, 127, 128, 129

—L—

LCAO (linear combination of atomic orbitals) approximation, 21

Lewis acid, 2, 50, 61, 84, 87, 99, 103, 119, 120, 121, 133, 136, 142, 143

Lewis and Brenstedt acids, 9, 23, 25, 31,
42, 44, 46, 49, 119, 125, 129, 139

—M—

magnesium chlorides, 84, 85, 86, 87, 113,
116, 119, 133, 136
Mendeleev's law, 30, 31
MINDO/3 method, 23, 40, 41, 52, 83, 105
molecular orbitals (MO), 21
Molecular system coordinates (MSC), 27,
28, 29

—O—

olefins, 1, 2, 3, 4, 5, 23, 37, 42, 44, 49, 50,
51, 52, 54, 55, 64, 65, 66, 67, 70, 71, 72,
81, 83, 84, 93, 103, 108, 109, 110, 111,
112, 116, 119, 120, 125, 128, 129, 132,
133, 136, 137, 138, 142, 143, 144, 145, 148
oxonium salts, 2

—P—

propylene, 51, 71, 81, 82, 83, 108, 111, 113,
119
proton detachment, 5, 31, 36, 44, 47, 48,
55, 61, 62, 63, 65, 69, 128, 130, 141
protonic acids, 2, 6

—Q—

quitones, 2

—T—

toluene, 59, 67

—V—

vinylic esters, 2

—Z—

Zero Differential Overlapping (ZDO), 22